What firefighters are saying...

To understand the art of firefighting, you have to understand physics, fire behavior, and human behavior. You also have to understand how humans, specifically firefighters, learn to apply acquired knowledge. This requires critical thinkers and intellectuals at the helm of the fire service, which is predominantly driven by trial, error, and experience—otherwise known as "tradition." Dr. Clark has been a trailblazer in critical thinking and applying intellectual "common-sense thought" to problems facing firefighters, even at the risk of enduring criticism for challenging traditional practices. Remaining confidently secure in his core values and perspectives truly makes him one of the great minds of the American Fire Service.

Captain Raul A. Angulo
Ladder Company 6, Seattle Fire Department

Dr. Clark's confrontational approach to line-of-duty deaths in the fire service is eye-opening. The idea that seatbelt use could keep my fellow firefighters' names off a memorial plaque challenged my mode of thinking. I was angry at first, but his force of logic and wellspring of true concern won me over. Thank you, Burt Clark!

Duane Hughes
*Chief of Training, Columbus
Fire and Rescue in Mississippi*

Throughout the history of the U.S. Fire Service, many giants have challenged the way we thought and operated, and drug us kicking and screaming toward progress. They challenged the norms and were not afraid to upset the status quo. Dr. Clark is without a doubt one of these giants. He has never shied away from asking the tough, taboo questions that we need to address in order to truly reduce firefighter injuries and fatalities. He has an unparalleled passion for the fire service and a true desire to make it safer.

Brent Batla
*Deputy Chief, MS EFO CFO,
Burleson Fire Department, in Texas*

Burt Clark has dedicated his professional life to teaching firefighters and working to reduce line-of-duty deaths. His matter-of-fact style forced me to take a good look at what kind of fire officer I was and what kind of fire officer I wanted to become. I am grateful for his passion and his perspective. Dr. Clark, thank you for sharing!

Lauren A. Johnson
*Section Chief, Aircraft Rescue Fire Fighting
Coordinator, Dallas Fire-Rescue Department*

THE AMERICAN FIRE CULTURE

Dr. Burton A. Clark EFO

I Can't Save You, But I'll Die Trying: The American Fire Culture
by Dr. Burton A. Clark EFO

©2015 Dr. Burton A. Clark EFO

Burt and Carolyn Clark conduct lectures and workshop on leadership and the American Fire Culture. They can be reached at drburtclark@gmail.com.

Published by PREMIUM PRESS AMERICA

All rights reserved. No part of this book may be reproduced or transmitted in any form or by any means, electronic or mechanical, including photocopying, recording, or by any information storage and retrieval system, without prior written permission of The Publisher, except where permitted by law.

ISBN 978-1-887654-57-9
Library of Congress Catalog Card Number 2014959152

PREMIUM PRESS AMERICA books are available at special discounts for premiums, sales, promotions, fundraising, or educational use. For details contact The Publisher, P.O. Box 58995, Nashville, TN 37205; or phone toll free 800-891-7323, or 615-353-7902, or fax 615-353-7905, or go to www.premiumpressamerica.com

Text and cover design by Armour&Armour www.armour-armour.com

Printed in the United States of America
10 9 8 7 6 5 4 3 2 1

ON THE COVER: The National Fallen Firefighters Memorial Park was erected in 1981 on the campus of the National Fire Academy in Emmitsburg, Maryland. The monument commemorates every line-of-duty deaths since its creation. Tradition holds that when a new death is documented, the firefighter's respective state flag along with the U.S. flag are presented at half-mast. The public is invited to visit. *Photo by Tony Burkett, President, South Newton Township Volunteer Fire Company, Pennsylvania*

Dedicated to the American fire service,
a noble calling

and

Carolyn Smith-Clark, a fire service leader in her own right, my wife, lover, friend, and soul mate who taught me how to love and be loved by children, grandchildren, and a great-granddaughter. The precious present is a gift we give to each other. I am a happy and blessed man because of her.

Contents

The Firefighter's Genes: Fast / Close / Wet / Risk / Injury / Death

'I Don't Want My Ears Burned'	21
Firefighting: Still Dangerous!	23
Firefighters Have to Get Killed; It's Part of the Job	29
Charleston Nine: Do the Right Thing or Look the Other Way	33
Your Behavior Comes from Ben Franklin's DNA: Fast, Close, Wet, Risk, Injury, Death	36
Fire/EMS Safety & Health Week: Rules vs. DNA	45
Why the Fire Service Should Avoid the Term 'Line-of-Duty Death'	51
Editorials Thirty-Seven Years Apart Agree: Getting Killed Is Not Part of the Job	58
The Ultimate Ethics of the Badge	63
Firefighter Philosophy In One Word: Why?	66

Seatbelts Save Lives: Why Don't We Wear Them?

To Be or Not To Be a Tattletale	73
How to Get Firefighters to Wear Seatbelts	76
Seatbelts: The Hugh Lee Newell Story	89
We Killed Firefighter Brian Hunton	96
Twelve Deaths This Year: It's Time for Seatbelt Hardball	101
Our Seatbelt Tale	107
The Princess, the Governor, and the Firefighter	110
LODDs: We Will Forget You!	112
Two Seatbelt LODDs Remembered	117
The Role of Leadership and Seatbelts	127
What Do I Stand For?	133
The Great Firefighter Seatbelt Lie	136

Mayday! Mayday! Mayday! Know When to Call It

Mayday! Mayday! Mayday! Do Firefighters Know When to Call It?	145
When Would You Call Mayday! Mayday! Mayday?	153
You Must Call Mayday for RIT to Work. Will You?	157
Calling A Mayday: The Drill	168

Best Practices

Operation Return: A Learning Experience	179
Ethnographic Interview of a Burn Patient	184
Lessons from America's Best-Run Fire Departments	189
Tests Are An Important Fire-Rescue Service Tool	206
Test Your Legacy Potential	209
Am I a Competent And Courageous Firefighter?	215
Train the Way You Fight—Fight the Way You Train	221
The Tale of Two Proactive Fire Chiefs	225
The White Elephant: Fire Department Response Time	228

The Pursuit of Higher Education

Feel the Door	233
Higher Education and Fire-Service Professionalism	238
Reading and Writing Equal Professionalism	247
The Best of the Best	250
Who Needs a PhD?	263
Getting a Doctoral Degree	271
Professor Frank Brannigan Taught Us More Than Building Construction	276

Speeches

What Matters	287
See the Light. Be the Light	290
Farewell to the National Fire Academy	304

Acronyms

AACFD	Anne Arundel County Fire Department
ARP	Advanced Research Project
ASTM	American Society for Testing and Materials
BLS	Bureau of Labor Statistics
CDL	Commercial Driver's License
CFOD	Chief Fire Officer Designee
CFPS	Certified Fire Protection Specialist
EFOP	Executive Fire Officer Program
EIB	Emergency Identifier Button
EKG	Electrocardiogram
EMS	Emergency Medical Service
EMT	Emergency Medical Technician
EMTB	Emergency Medical Technician Basic
EMTP	Emergency Medical Technician Paramedic
FAMA	Fire Apparatus Manufacturers Association
FDIC	Fire Department Instructors Conference
FDNY	Fire Department City of New York
FLSA	Fair Labor Standards Act
GPM	Gallons Per Minute
HAZMAT	Hazardous Materials
IAFC	International Association of Fire Chiefs
IAFF	International Association of Fire Fighters
ICS	Incident Command System
IDLH	Imminent Danger to Life and Health
KIA	Killed in Action
LODD	Line-of-Duty Death
LUNAR	Location, Unit Number, Name, Assignment, Resources
MFRI	Maryland Fire Rescue Institute
NFA	National Fire Academy
NFFF	National Fallen Firefighters Foundation
NFPA	National Fire Protection Association
NHTSA	National Highway Traffic Safety Administration
NICU	Neurological Intensive Care Unit
NIMS	National Incident Management System
NIOSH	National Institute for Occupational Safety and Health
NIST	National Institute of Standards & Technology
OJT	On-the-Job Training
ORA	Outstanding Research Award
OSHA	Occupational Safety & Health Administration
PASS	Personal Alert Safety System
PPE	Personal Protection Equipment
SCBA	Self-Contained Breathing Apparatus
SOP	Standard Operating Procedure
UL	Underwriters Laboratory
USFA	United States Fire Administration
VFD	Volunteer Fire Department
VFIS	Volunteer Firemen's Insurance Services

Foreword

THE ART OF fighting fires has long been a topic of many accomplished authors. They teach skills needed to do the job, but is that enough? Anyone who has listened to a Burt Clark lecture or read one of his articles will know that IT IS NOT ENOUGH. Burt has made it his life's work to examine the culture, test the theories, and document ways firefighters die needlessly. Saving lives is not only what he does but who he is and finding ways to keep firefighters safe is his passion.

Burt causes us to take stock of our lives, what we want to do, and how we do it. There is no news coverage of firefighters doing the right thing, nobody give awards for it…so why does it matter? Ask any mother, wife, or child who experienced a line of duty death. To know that their firefighters died simply because they did not buckle a seat belt, or a chin strap, is beyond devastating.

I will always wonder had this book been published six years ago, would Robin still be alive? We have to make this commitment for our own families. Robin's girls, Sierra and Courtney, know the cost of not doing so.

—Arlene Zang, FF/P, proud mother of
Captain Robin Broxterman,
Colerain Township, Ohio
LODD 04/04/08

Preface

AT MY FIRST FIRE I almost killed another firefighter. I was ashamed of myself, sick in the stomach because I didn't know what I was doing and I was dangerous. Two years later I graduated with the highest academic score from the District of Columbia Fire Department Recruit Class 249, because the fire service is a life-and-death occupation with no room for error.

You believe firefighters will come save you, your children, and your property if there is a fire. Firefighters also believe they can save you and have been getting injured and killed trying to do so from Ben Franklin's time to today.

In February 1974 Laurel Volunteer Fire Department in Maryland responded to seven civilian fire deaths in three home fires; all the victims were dead before the alarms were received. I felt helpless because all of our equipment, training, skill, and macho could not save them. There had to be a better way, so I became chair of the Fire Prevention Committee.

In 1975 the Laurel Volunteer Fire Department received the Maryland State Firemen's Association's Fire Prevention award for its fire safety home inspection and home smoke detector campaign. In 1976 the Washington, D.C., Mayor's Office asked me, a D.C. firefighter, to help the city address the fire problem. The cities campaign was initiated as a result of several fatal fires in neighborhoods where the closest fire companies had been closed due to rolling brown-outs from budget cuts. Washington, D.C., was the first city to have a mandatory smoke-detector ordinance for all existing and new residential occupancies and a training program for all firefighters on how to educate the public about smoke detectors.

In 1978 I was detailed to the National Fire Academy to develop and conduct the Smoke Detector Training Program.

I Can't Save You But I'll Die Trying: American Fire Culture

This course was created to help fire departments nationwide implement campaigns to promote the installation of residential smoke detectors. Smoke alarms have helped reduce fire deaths nationwide. Today, when there is a home fire death most of the time there are no working smoke alarms. When I visit my children and grandchildren I check the smoke alarms. Sometimes I have to change the batteries or install new smoke alarms. Our smoke alarm work is not done.

The purpose of the National Fire Academy is to "advance the professional development of fire service personnel and other persons engages in fire prevention and control activities" (PUBLIC LAW 93-498-OCT. 29, 1974). This is a purpose I committed most of my adult career to. The NFA is our Harvard, West Point, and Top Gun schools all rolled into one.

When the National Fallen Firefighter Memorial was built in 1981, the importance of the NFA's purpose became evident and visceral to me. As I drove onto campuses each morning, most of the time the flags were at half-staff indicating another firefighter had been lost. There are 3,838 (1981 to 2014) names on the memorial. About a hundred names are added each year except for 343 names added as a result of 9/11.

Firefighter Brian Hunton is one of those names. He was a National Fire Academy graduate who fell out of his firetruck in 2005 on the way to a house fire; he did not have his seatbelt on. I cried and felt ashamed. We have a hard time getting firefighters to buckle their seatbelts because they believe seatbelts will slow them down and they will not be able to save civilians. Some states even exempt firefighters from using seat belts.

Captain Robin Broxtermen was scheduled to attend my NFA course. When I leaned of her loss I cried. Her fire department lent me a helmet; Robin attended the class for two weeks posthumously and received her National Fire Academy certificate.

From the beginning of my career I have rejected the philosophy that firefighter injury and death is part of the job. And

Burton A. Clark

I have feared the fact that citizens rely on me to come save them from fire when I know I will fall short of that expectation. We all must do our best when it comes to fire safety—and we can all do better.

This book is about the journey to answer my fire-service calling. After forty-five years of learning what to do next and trying to do better, I hope this book helps save the lives of my neighbors and my firefighting family.

–Dr. Burton A. Clark EFO

NOTE: A portion of the proceeds from the sale of this book goes to the Firefighters Cancer Support Network and the National Fire Heritage Center. Please make individual and organizational contributions directly to these groups.

The Firefighter Cancer Support Network is a 501(c)3 nonprofit organization established in 2005 by Los Angeles County Firefighter Paramedic Mike Dubron, a survivor of stage IV colon cancer. Today, FCSN's objective is to provide timely assistance to all fire and EMS personnel and their family members who have been diagnosed with cancer. FCSN also develops and delivers cancer awareness, prevention, education, and outreach programs nationwide. FCSN maintains a unique network of more than 120 mentors with personal experience facing a multitude of cancers; it provides one-on-one support for firefighters coping with cancer. FCSN relies upon hundreds of other volunteers in thirty-four states to provide vital services across America. FCSN does not provide legal or medical advice, but it does offer additional assistance with cancer-specific support programs, behavioral health services, and fire-service chaplains. FirefighterCancerSupport.org

The National Fire Heritage Center is a 501(c)3 organization created to preserve and protect the history of the American fire protection services and allied disciplines through identification, acquisition, restoration, and conservation of historic information. NFHC is committed to the preservation of ideas including the legacy of written and original works by individual contributors within the profession, the published history and literature of the American fire service industry, and the conservation of oral histories. It aims to establish a facility that will facilitate historical review and research to approaches and methods to save human life from destructive fires and mitigate fire losses in the future. NationalFireHeritageCenter.org

Introduction

DO YOU KNOW there are 1.1 million firefighters and 32,000 fire departments in the United States of America? About eighty percent are volunteers, and twenty percent are paid. More Americans die from fire then all other natural and man-made disasters combined annually, and firefighting is one of the most hazardous occupations. If you are a civilian reading these statistics you might not know this information; if you are a firefighter you will. Each reader will have a preconceived notion of what firefighters are, what they do, and why they do it. Your perceptions may change after reading this book. The individual articles came out of my experiences with specific incidents that compelled me to look deeper at myself and the fire service discipline that I love.

After forty-five years in the fire service, I have concluded, "I can't save you, but I will die trying!" I come to this reality as a result of being part of the American Fire Culture for almost five decades. As such I can trace my fire-service roots to Ben Franklin; my father and mother; the Kentland, District of Columbia, Laurel, and Mt. Airy fire departments; and the National Fire Academy. My journey is reflected in the forty-five essays in this book. Each is the result of some significant emotional event I experienced, what I learned, and my attempt to influence the readers and my fire-service colleagues.

If there is such a thing as destiny, I was destined to be a firefighter before I was born. My father was a fireman in the US Army in 1941, stationed at Pearl Harbor, Hawaii. One of my first toys (I still have it) was a ladder truck I received at three or four years of age. Today that toy is an antique and considering I just turned sixty-five years old and just retired, I may be in the same category.

Becoming a firefighter was destiny; being a writer was not. Some even thought it would be impossible for a child who initially could not read and write. Today, I would be classified as a learning-disabled student. When I was in school they just told my mother that I was lazy; Mom did not believe them. I will make sure my schools get a copy of this book for their libraries.

The universities that admitted a below-average student deserve credit for helping me work at being a scholar. Thank you Strayer, Montgomery, Catholic, Nova Southeastern, and John Hopkins; I am grateful to all my professors.

The fire service has been good to me. It has given me all I need according to Maslow's hierarchy of needs: physiological, safety, love/belonging, esteem, and self-actualization. The fire service has given meaning to my life. Putting this collection of essays together in a book is the continuation of my fire-service journey. It is a tribute and a way for me to say thank you to all the individuals and organizations I have interacted with along the way.

I hope this book helps you think, feel, and learn more about the American Fire Culture from one firefighter's experiences. Whether you are a firefighter or civilian you are an important part of our fire culture—past, present, and future—because lives are at stake.

NOTE: The first 40 essays in the book were selected by four fire service friends: Captain Raul Angulo, Seattle Fire Department; Battalion Chief Brent Batla, Burleson Fire Department in Texas; Battalion Chief Duane Hughes, Columbus Fire Department in Mississippi; and Battalion Chief Lauren Johnson, Dallas Fire Department. They were so kind to read all of my work and vote on what they believed to be the best. The last five essays are some of my most recent thoughts.

ns:
Fast / Close / Wet / Risk / Injury / Death

'I Don't Want My Ears Burned'
1976

IF THERE IS one thing common to all fire departments in the world, it must be bull sessions between emergencies. During a recent session, I was defending the importance of an academic education for the firefighter. Needless to say, I was alone in my opinion. The majority argued, "You don't need a college degree to ride on the back step." I countered with, "If we become more educated, maybe we'll kill fewer firefighters."

The next statement I heard frightened me. My lieutenant said: "Firefighters have to get killed; it's part of the job."

Do fire service personnel believe that being injured or killed is part of the job? Maybe we do, because the number of casualties keeps rising. Our military background may be partly to blame. In every battle there is an acceptable casualty level. What is the fire service's acceptable casualty level?

That which a society or group accepts is what it is likely to get. Polio became unacceptable; today it is almost non-existent. The commercial airline industry decided in the beginning that zero accidents would be the only acceptable level. That's why air travel is so safe. The fire service will stop being the most hazardous occupation only when we, its members, want it to.

Zero casualties won't come easily. We must become our own toughest critics. Every time an accident occurs we must determine who or what is at fault. No one easily admits that he made a mistake, but that's how we learn.

Most of us have had our ears burned at least once. Why?

Does your helmet have ear flaps? Were you taught to pull your ear flaps down? Did you forget to pull your ear flaps down? Did the ear flaps fail to protect you? Why were you so close to the fire? Firefighters get burned; it's part of the job.

If your answer is the last one, your ears will get burned again. Whether it is your ears getting burned, or another firefighter getting killed, it is someone or something's fault, not just a part of the job.

I know what some readers will be thinking: "If you can't take it, Clark, you should get out of the service." This attitude is one reason that we are number one in hazardous occupations. As a member of the "new breed," I think zero is the only acceptable level of casualties.

This idea evolved during a bull session. Maybe my whole theory is bull. I hope not.

Firefighting: Still Dangerous!
2003

FIREFIGHTING IS dangerous. We intuitively believe this, but it is not true according to a recently published article that states "firefighting as an occupation does not have as many fatalities as other occupations (Peterson, 2002, p.1). Peterson's conclusion is based in part on a U.S. Department of Labor Bureau of Labor Statistics study that included firefighter fatalities from 1992 to 1997 (Clarke and Zak, 1999). When Peterson used the same statistics procedure on the year 2000 firefighter date, he calculated that "firefighting is not even in the top fifteen occupations in respect to risk of fatal injury" (2002, p. 2).

When we read the articles we knew they were wrong in our gut, but the research and statistical facts proved them to be correct. Further examination of the research methodology revealed that the statistical foundation of these conclusions was wrong. The fundamental error is based on the assumption that firefighting is just like any other occupation and therefore can be directly compared to other occupations using the same normalization procedure.

The Peterson article and the BLS study compared death

rates per 100,000 employed workers. They derived their output by dividing the total number of fatalities in each occupation by the total number of employed workers. The output was computed by multiplying 100,000 to arrive at the fatality rate per 100,000 workers. What this method assumes is that 100,000 workers in each occupation are equalized and the fatality rate per 100,000 employees can be reported.

With this statistic each occupation can be compared to the other. This method of calculating risk assumes that all occupations are equally at risk. But the procedure does not define "at risk." For example, firefighters have a fatality rate of 18.3 per 100,000 workers and roofers have a fatality rate of 27.5 per 100,000 workers (Clarke and Zak, 1999). When comparing these two fatality rates, it appears roofers are at a higher risk of death on the job than firefighters.

Fatality rate for 100,000

Occupation/Industry	Employment	Index of Relative Risk
All occupations	4.7	1.0
Timber cutting/logging	128.7	27.4
Fishers	123.4	28.3
Water transportation	94.2	20.0
Aircraft pilots	83.3	17.7
Construction laborers	41.1	8.7
Truck drivers	27.9	5.9
Roofers	27.5	5.9
Farming	27.5	5.9
Firefighters	18.3	3.9
Police	14.0	3.0

The critical departure we take from previous firefighter fatality studies is trying to determine when firefighters or other workers are actually at risk. This question may seem ridiculous, but stay with us because answering it will change how firefighter injury and fatality statistics are analyzed.

Is a firefighter at risk in the station washing the firetruck? Is a firefighter at risk eating dinner or sleeping at the station? Is a volunteer firefighter at risk while waiting at home for an emergency call? The answer is *no*. When is a roofer at risk? This answer is much easier. A roofer is at risk when they are engaged in the job of roofing. So the new concept we are presenting is that "time at risk" of injury or death must be considered when determining the risk factor of any occupation.

Both of us have nailed shingles to a roof at least once, but we do not consider ourselves roofers. We apologize to all roofers if we get these next calculations wrong, but we are attempting to illustrate our concept. If I am a roofer I am probably engaged in roofing activities—i.e., on the roof, climbing, carrying, cutting, or hammering—six hours out of every eight-hour day of work. Based on this assumption a roofer is at risk six hours out of every eight-hour workday. Now if our friend the roofer works six days a week (we gave him eight hours' overtime) in a forty-eight-hour workweek the roofer is at risk (engaged in doing the job) thirty-six hours.

The hard part of this new concept is figuring out firefighter "time at risk." Is a firefighter at risk on an EMS call? That is a FLSA question, which is beyond this article. Our firefighter assumption is that firefighters are "at risk" whenever they are on an emergency call. This time period commences when they leave the station, lights and siren, and continues until they return to the station, regardless of the type of emergency. Now how do we calculate this considering all the different fire departments from the New York City Fire Department to the Laurel Volunteer Fire Department?

We chose two extremes we have some connection to, Engine 10 in the District of Columbia Fire Department, Washington, D.C. (the busiest engine company in the United States) and the Vigilant Hose Volunteer Fire Department, Emmitsburg, Maryland. (It protects the National Fire Academy.) What is the connection? We attended LODD memorial services for firefighters from both companies. (We dedicate this article to firefighters Terry L. Myers VHVFD, LODD February 15, 1999, and Anthony S. Phillips, Engine 10, DCFD, LODD May 30, 1999.)

We randomly chose a forty-eight-hour period from 0800 hr January 30 to 0800 hr February 1, 2003, and calculated the number of hours the companies were at risk. Engine 10 had a total of thirty-seven responses equaling ten hours of time at risk. The VHVFD had one response equaling one hour of time at risk. The average of the two is 5½ hours (11 divided by 2), which we rounded it to an even 5.0. We can now use "time at risk" to compare the two occupations. In a forty-eight-hour workweek a roofer is at risk thirty-six hours and a firefighter is at risk five hours. The "time at risk" of the firefighter is one-seventh (5/36=7.2) that of the roofer or any other occupation engaged in job duties six hours out of each eight-hour day of work.

Now lets go back to our risk factors from the Clarke and Zak article. Roofers have a risk factor of 27.5 with thirty-six hours of time at risk. Firefighters have a risk factor of 18.3 with five hours of time at risk. To norm the time at risk for each occupation we need to multiply 18.3 by 7 to get equal hours of time at risk with the roofers (18.3 x 7 = 128). Firefighter's risk factor per one hundred thousand workers adjusted for time at risk is 128. This puts firefighters at the top of the fatality risk list, equal to timber cutters/logging at 128.

Another interpretation error in the Peterson article is the raw firefighter death counts. In the year 2000 firefighters had 102 LODD, miners had 156, manufacturing had 670. So it looks as if firefighting is less dangerous than mining or manufacturing. Again we need to figure out how to norm the work done by each job. Work output is one calculation method. In 2000 we had 1,708,000 fires so we can look at the number of deaths per fire: 102 divided by 1,708,000 or one firefighter LODD for every 16,745 fires (1/16,745).

Now these fires can range from a four-alarm blaze, a car fire, food on the stove, or a trashcan; for our concept it does not matter. Mining is reported in tons of material. So 156 deaths divided by 6,493,600,000 tons (the amount of material mined in the U.S. in 2000) or 1/4,162,564. The manufacturing occupation recorded 670 deaths with 230,819,501 units* or 1/689,013. Now, based on our research, firefighting can be reported as forty-one times more deadly than manufacturing and 248 times more deadly than mining.

The fundamental statistical problem stems from trying to compare firefighting to other occupations, using standard statistical procedures. Standard statistical procedures misrepresent firefighter death and injury statistics. Another major flaw is that volunteer firefighters are not included in some workforce studies, including the BLS article from Clarke and Zak. Only paid firefighters are counted. The fire service needs to be studied under its unique conditions and research boundaries. Time at risk is a critical factor in the fire service that must be included in any comparison to other occupations.

Finally, Peterson's statement, "Unfortunately, there will always be firefighter deaths because of the dynamic work of fire and related emergencies" (2002, p. 4) is as unacceptable as the statement made in 1976 by a company officer: "Firefighters have to get killed; it is part of the job." (See page 29.) In comparison, the airline industry has no acceptable death rate. This is one reason flying is so safe. If one airline passenger was killed every 16,745 flights no one would fly.

The fire service is the most dangerous occupation, career and volunteer, in the U.S. This is a reality. We cannot let statistics tell us differently. The death or injury of a firefighter is more than a number. It is a tragedy that we can prevent. Will we? Will you?

This article was co-written by David M. Ballard.

*Manufacturing unit is identified for this article as the total number of cars produced in the U.S. in the year 2000 (9,527,501) and tons of steel produced in the U.S. in 2000 (221,292,000).

References

Peterson, David F. (2002) Firefighting: Risky business. Dealing with the risk of emergency response. Firehouse.com

Clarke, Cindy, and Zak, Mark J. (Summer 1999) Fatalities to law enforcement and firefighters, 1992-97. Office of Safety, Health, and Working Conditions, Bureau of Labor Statistics.

Clark, Burton A. "I don't want my ears burned." Fire Command (July 1976, p. 17, formerly published by the National Fire Protection Association)

Firefighters Have to Get Killed; It's Part of the Job

2008

D O YOU AGREE or disagree with this statement? Does the fire service and society believe that?

The major national fire service organizations: NFPA, IAFC, NVFC, NFFF, IAFF, USFA, plus OSHA, NIOSH, and NIST have produced thousands of firefighter safety standards, laws, training material, reports, and investigations over the past thirty-one years. National fire publications, conferences, and campaigns have focused on reducing firefighter injuries and deaths. The federal government mandates ICS training, SCBA testing and training, and NIMS compliance. As a result of 9/11 more federal grant money for apparatus, PPE, equipment, and training has been spent than ever before. State and local fire service organizations conduct countless hours of training and spend millions of dollars on firefighter safety every year.

In 1976, there were 107 line of duty deaths. In 2007, we had 115 line of duty deaths and about eighty thousand injuries.

The number of deaths and injuries has not changed much over the past thirty years. The reasons for the deaths and injuries have not changed. Because, in spite of all the safety talk and safety programs our safety belief, attitude, and behavior has not changed.

The NIOSH line-of-duty death studies report that we do not follow our own safety SOPs, national standards, and training doctrine. We do not use our safety equipment. We do not hold firefighters, officers, or chiefs responsible and accountable when it comes to safety. More fire-service personnel are disciplined for being late for work than safety violations. (I don't have the data to prove this, but no one has ever disagreed or presented information to the contrary.) In other words, we tolerate and accept safety misconduct, which can and does result in firefighter death or injury.

In 2004, Martha Stewart was convicted of lying to investigators about a $228,000 stock sale. She served five months in prison and paid a thirty-thousand-dollar fine. The government was sending a message that corporate misconduct would not be tolerated.

In 2007, the City of Los Angeles awarded a firefighter $2.7 million to settle a lawsuit because his brother firefighters put dog food in his spaghetti. Racial misconduct will not be tolerated in the fire service. Societies know how to hold individuals and organizations responsible for behavior that will not be tolerated.

In Charleston, South Carolina, the mayor reported the fire was, "beyond the firefighting capability of any fire department." The fire chief concluded that nothing could have been done differently. South Carolina state regulators found several workplace safety violations at the fire that killed nine firefighters. The city paid three thousand dollars in penalties while not admitting any wrongdoing ("Scrutiny Feeds Firehouse Tensions After Fatal Blaze," Adam Hochberg, National Public Radio, January 16, 2008).

In 2007, twelve firefighters were killed in the line of duty without their seatbelts on; three hundred have died in the past thirty years. There has been no discipline or penalties or accountability for their deaths.

Our belief, attitude, and behavior reinforce the lieutenant who said "Firefighters have to get killed; it's part of the job." At the time he did not clarify if he meant firefighter deaths due to heart attack, vehicle crash, getting hit by a vehicle, building collapse, falling through a floor or roof, flashover or backdraft, getting lost or trapped, running out of air, trying to rescue a victim, or falling from a ladder, roof, or apparatus. You will have to decide which line-of-duty deaths are part of the job. There is little or no accountability, responsibility, or discipline for firefighter death or injury from the fire service, elected officials, or government agencies. There are no consequences for safety misconduct. Because, we have convinced others and ourselves that firefighter death and injury is just part of the job.

Take the test (circle one): Firefighters have to get killed; it's part of the job. • Agree • Disagree

I disagreed in 1976 and I still disagree today. When a firefighter is killed or injured, something went wrong. It is not part of the job. We can and must behave differently. We do not need to learn anything new, get new equipment, or have new SOPs and standards. We must do every task one hundred percent correctly one hundred percent of the time. When we do not meet this goal there must be accountability and responsibility followed by discipline. We know how to fix misconduct at the organizational and individual level. When a NIOSH firefighter line-of-duty death report reads:

- All personnel followed all SOPs and training doctrine;
- The structure met all fire codes;
- Nothing could have been done differently;
- Nothing can be learned from this firefighter's death;
- This line-of-duty death was part of the job;

then the lieutenant, chief, and mayor will be correct. Until then, be afraid, be very afraid. Because someone near you or in your chain of command believes that firefighters' getting killed is part of the job.

I know what some readers are thinking. "Clark is full of bull; he is just an old guy who can't take it anymore." You are right. When enough of us can't take death and injury as just part of the job, death and injury on the job will stop. I hope it doesn't take thirty more years.

Author's note

I am grateful to the January 2008 NFA Fire Service Communications class and instructors Paul Burkhart and Mike Chiaramonte for editing this article. I am grateful to the February 2008 NFA Organizational Theory in Practice class and instructors Mac Greenland and Al Thompson for a philosophical critique of this article.

Charleston Nine: Do the Right Thing or Look the Other Way
2008

COULD YOUR FIRE department lose nine firefighters in a building fire tonight? Could some other fire department in your state lose firefighters in a building fire tonight? Could some fire department somewhere in the United States lose nine firefighters in a building fire tonight?

The answer is *yes* to one or more of these questions.

Will the cause of these firefighters' line-of-duty deaths be something different than the contributing factors in the loss of the Charleston Nine? Have we killed firefighters in any new ways, excluding 9/11, in the past thirty years? Has every fire department in the country corrected all the deficiencies identified in Charleston?

The answer is *no* to one or more of these questions.

A fire chief retiring or a report being released will not change the *yes* to a *no* or the *no* to a *yes*.

Every firefighter, ever fire officer, and every chief officer must identify what they will *start doing* today and stop doing today so they, their crew, or their department will not be one of the next nine firefighters to be killed in a building fire.

It is never about the other person's doing the right thing. It is always about my doing the right thing. I hope my partner, officer, incident commander, and chief help me because I am human.

Have you ever seen the deficiencies identified at the Charleston fire repeated at other fires? What did you do about it? Or did you choose to look the other way?

This poem helps give me the courage to do the right thing:

I Chose to Look the Other Way
I could have saved a life that day,
But I chose to look the other way.
It wasn't that I didn't care,
I had the time, and I was there.
But I didn't want to seem a fool,
Or argue over a safety rule.
I knew he'd done the job before,
If I called it wrong, he might get sore.
The chances didn't seem that bad,
I've done the same, he knew I had.
So I shook my head and walked on by,
He knew the risks as well as I.
He took a chance, I closed an eye,
And with that act I let him die.
I should have saved a life that day,
But I chose to look the other way.
Now every time I see his wife,
I'll know I should have saved his life.
That guilt is something I must bear,
But it isn't something you need to share.
If you see a risk that others take,
That puts their health or life at stake,
The question asked, or the thing you say,

> Could help them live another day.
> If you see a risk and walk away,
> Then hope you never have to say,
> I could have saved a life that day,
> But I chose to look the other way.
>
> —Don Merrell, August 2003

The fire service looked the other way before the Charleston fire. We have to live with that loss and bear the guilt. Will we all do the right thing? From now on? If not, what will be our excuse the next time and the next time—and the next time?

Training, equipment, SOPs, standards, codes, incident command systems, and ratings have not solved our problem. Maybe we should try poetry.

Your Behavior Comes from Ben Franklin's DNA: Fast, Close, Wet, Risk, Injury, Death
2011

WHEN I WAS a rookie fireman in 1970 at the Kentland Volunteer Fire Department Company 33, Prince George's County Fire Department in Maryland, an old-timer (he was thirty-five, I was twenty) told me, "The next call you go on may be the biggest fire in your career, so you must be ready." At the time, the KVFD was responding to about a thousand alarms per year, and many were working fires. This advice was burned into me, literally and figuratively, at the affective, cognitive, and psychomotor levels of learning. I was among the top ten responders my first year as a fireman, and I was injected with Ben Franklin's DNA for the next forty years.

The 2011 safety-stand-down-week theme was "Surviving the Fireground." When does fireground survival begin? The place to start ensuring your survival on the fireground is at the fire station before the alarm. If there is no water in the engine's tank, or your self-contained breathing apparatus has only one thousand PSI, or your hood and gloves are missing from your bunker coat, or the battery in your portable radio is dead, the chances of your surviving the fireground are beginning to diminish.

When I read the NIOSH line-of-duty-death reports on fireground fatalities, I wonder what little change in the sequence of events could have avoided the error that lead to the tragedy? We have not invented any new way to injure or kill firefighters; Professor Frank Brannigan taught me that in 1974, and it is still true today. (See page 276.)

If you do not put your seatbelt on before the apparatus begins to move, or if your driver fails to check that everyone is buckled in and your officer fails to enforce the seatbelt standard operating procedure and your chief doesn't consider seatbelt use a priority, the chances of you and the crew's getting injured or killed on the fireground go up because you are not ready for the biggest fire of your career—or any fire for that matter.

It seems as if you, your team, and your fire department have decided that SOPs, safety equipment, duty, and accountability do not apply to you. All the personnel in your department can pick and choose what they do and don't do. If you and your crew are like this, you are in the majority of the fire service today.

Blame it on Ben

Why does this persist? We learned it from Ben Franklin over the past 275 years. I know you are thinking, "Clark has finally lost his mind. He is blaming seatbelt and fireground LODDs on Ben Franklin." I am not alone in this thinking, so keep reading.

The number-one firefighter life-safety initiative from the National Fallen Firefighters Foundation's "Everyone Goes Home" campaign states, "Define and advocate the need for a cultural change within the fire service relating to safety; incorporating leadership, management, supervision, accountability, and personal responsibility."

We have to define fire-service culture before we can change it. Having a common definition of anything is not easy. Professionals use common definitions among themselves so they can clearly communicate with each other. For example, if I tell you the patient has an open fracture of the left femur, you could all draw a picture of what it looks like, whether you are an MD, EMTP, EMTB, or first responder.

However, if I asked you to draw a picture of an engine company, we would get a bunch of different drawings. Having a shared definition of fire-service culture is difficult because the words must mean the same thing for all fire departments, regardless of size or location. What is the definition of fire-service culture? What does it look like? Do we all draw the same picture of fire-service culture? Does a culture exist in 1.2 million firefighters and thirty-two thousand fire departments nationwide? Let's start by defining culture as it relates to organizations.

Organizational culture

In *Organizational Culture and Leadership* (2004, Edgar H. Schein, Professor Emeritus at MIT), the author defines culture as, "A pattern of shared basic assumptions that was learned by a group as it solved problems of external adaptation and internal integration, that has worked well enough to be considered valid and, therefore, to be taught to new members as the correct way to perceive, think, and feel in relation to those problems." Firefighter translation: "Why we do what we do."

Schein also tells us that occupations can have a shared culture if the following conditions are present: intense period of education and apprenticeship, reinforcement of assumptions at meetings, and continuing education sessions. The practice of the occupation requires teamwork and reliance on peer-group evaluation, which preserves and protects the culture. The fire service meets these conditions, so the notion that fire-service discipline has a shared culture is reasonable. This supports the NFFF reference to the need to change the fire-service culture. Before we can change the culture, we have to be able to identify what the culture is. Schein explains that culture has three levels.

Artifacts: Visible organizational structures and processes. These are the things we can see, touch, and read.

Espoused beliefs and values: Represented by our strategies, goals, and philosophies (espoused justifications). This is what we tell each other and the public what we do, how we do it, and why we do it.

Underlying assumptions: Taken-for-granted beliefs, perceptions, thoughts, and feelings (ultimate source of values and action). Schein refers to this culture level as the DNA of an organization. For the fire service, this is the basic DNA of what it means to be a firefighter. This genetic code has been passed down from generation to generation of firefighter over the past 275 years. If you do not have it, you are not a real firefighter. This is where Benjamin Franklin started the genetic pool we have today. According to Wikipedia, "A gene is a unit of heredity and is a region of DNA that influences a particular characteristic in an organism."

It's in our genes

In a paper entitled "FAST/CLOSE/WET" delivered at a Public Entity Risk Institute symposium about "Reducing Firefighter Deaths and Injuries: Changes in Concept, Policy and Practice," Chief Allen V. Brunacini identified the first three genes of a firefighter's DNA. Brunacini wrote, "Ben [Franklin] realized that when there was a fire that the situation required rapid response, so he taught his fire lads that they must be FAST. He also knew that he did not have long-range hydraulic application equipment, so his firefighters had to get

CLOSE to the fire. Ben also understood that the fire could not live in the same space with an adequate amount of water so he told his troop get the fire WET."

The next three genes, RISK/INJURY/DEATH, are all part of the human experience with uncontrolled fire. Human beings have been at risk of uncontrolled fire, injury by fire, and death by fire from the beginning of time. Our bodies cannot live in the heat, gases, and oxygen-depleted environments that fire can create. Our environment, property, and possessions can be destroyed by fire. Anyone who tries to manually control an unwanted fire or save someone or something in the path of an unwanted fire puts themselves at great risk, which can lead to injury and death. Ben Franklin knew this, so his firefighters had to accept this as part of what it meant to be a firefighter. The citizens knew this, so they held the firefighters in great esteem because when called for help, the firefighters would put their bodies between the fire and the citizen to save and protect individuals, families, property, and communities from the ravages of fire.

The fire service and society today continue to consider RISK/INJURY/DEATH part of the characteristics that exist when humans get in the path of uncontrolled fire. Recently, this idea was supported by an analysis of NIOSH LODD reports, which helps to identify the cultural paradigm of firefighting and the public image of the fire service.

Doctors Kunadharaju, Smith, and DeJoy from the College of Public Health at the University of Georgia published a paper titled "Line of Duty Deaths among U.S. Firefighters: An Analysis of Fatality Investigations." They studied 189 NIOSH reports that included 213 LODDs from 2004 to 2009. The NIOSH reports made a total of 1,167 recommendations to reduce firefighter injury and death. The researchers categorized the recommendations into five factors: Incident Command, Personnel, Equipment, Operations/Tactics, and External. The researchers applied root-cause analysis techniques to the data set to determine the basic or higher-order causes that they classified as under-resourcing, inadequate preparation for/anticipation of adverse events, incomplete adoption of incident-command procedures, and sub-optimal personnel readiness. An important point they make is that these higher order causes "do not provide any definitive insights as to their origin," but "may actually be tapping the basic culture of firefighting." The researchers go on to make the following comment about the core culture of firefighting:

> Operating with too few resources, compromising certain roles and functions, skipping or short-changing operational steps and safeguards, and relying on extreme individual efforts and heroics may reflect the cultural paradigm of firefighting. This should not be construed to be a culture of negligence or incompetence, but rather a culture of longstanding acceptance and tradition. Within many fire-service organizations, these operational tenets may be accepted as "the way we do things." Moreover, this tolerance of risk may be reinforced both externally and internally through the positive public image of firefighters and firefighting and internally through the fire service's own traditions and member socialization.

Chief Brunacini confirms these comments from Kunadharaju, Smith, and DeJou with the following statement, as only he can, in firefighter language:

> When the fire kills us, our department typically conducts a huge ritualistic funeral ceremony, engraves our name on the honor wall, and makes us an eternal hero. Every LODD gets the same terminal ritual regardless if the firefighter was taking

an appropriate risk to protect a savable life or was recreationally freelancing in a clearly defensive place. A fire chief would commit instant occupational suicide by saying that the reason everyone is here today in their dress blues is because the dearly departed failed to follow the department safety plan. Genuine bravery and terminal stupidity both get the same eulogy. Our young firefighters are motivated and inspired to attack even harder by the ceremonialization of our battleground deaths.

For the past 275 years, fire service DNA has been made up of these six firefighter genes FAST, CLOSE, WET, RISK, INJURY, DEATH (FCWRID). These are the underlying assumptions that are taken-for-granted beliefs, perceptions, thoughts, and feelings and are the ultimate source of values and action.

The entire fire-service discipline and general public use the FCWRID gene sequence or combination of the genes to predict, justify, explain, accept, reward, and improve the fire service. Before you all tar and feather me, or burn me in effigy, remember we and the general public do not do this consciously from malice or incompetence. We learned it from our ancestors who were doing the best they could at the time.

FAST thinking

I will use just one firefighter gene, FAST, to illustrate how it influences all levels of fire-service culture and our behavior.

Artifact: Lights and sirens, Opticom, response time standards, state and federal laws that exempt seatbelt use by firefighters, running to the apparatus when a building fire is announced, political discourse related to closing fire stations and increased response time: "If we close these fire stations our average response time will go from four minutes and forty seconds to five minutes and ten seconds."

Espoused beliefs and values: Closing a fire station puts the public at risk because we will not be FAST; if I put my seatbelt on it will slow me down; I can't put my seatbelt on with all my bunker gear on; my bailout equipment keeps me from putting on my seatbelt; no one beats us into our first-due area; no one steals our fire; firefighter safety is important; we have SOPs; the company officer did not have the time to look at side charley before entering the front door because the fast attack was used; the officer left his portable radio on the firetruck; the crew fell through the floor; no Mayday was called; the CO and firefighter died in the basement fire making the ultimate sacrifice.

Underlying assumption: I must be FAST. One of the worst things that can happen is for another fire company to beat you into your first-due area. A fire chief told me, "If we did not respond with lights and siren on all calls, we would not be an emergency service." Citizens will say, "It took the fire department a long time to get here." Get in there and get the fire; no one steals our fire. Firefighters get injured and killed responding to alarms in vehicle crashes without their seatbelts on. This is considered a line-of-duty death with full ceremonial honors at the funeral, community-wide shared grief, and LODD cash benefits from local, state, and federal levels.

When there is a firefighter LODD, the root cause is rarely, if ever, a technical problem. The underlying cause can be traced back to one or more firefighter genes that drive our behaviors resulting in the ultimate loss. We have accepted this for the past 275 years. If we continue to justify our behavior based on our firefighter genes, more of us will be injured and killed.

Benjamin Franklin was a scientist, inventor, printer, author, statesman, businessman, and educator. As a founding father of the American fire service, Ben would be disappointed in us today if all we inherited from him was FAST, CLOSE, WET, RISK, INJURY, DEATH.

Changing your DNA and genes is difficult, but you can change your behavior if you choose. Why should you change? Because "The next call you go on may be the biggest fire in your career, so you must be ready, if you want to survive."

You and I cannot change fire-service culture. But, as a firefighter, what one behavior can you change? As the apparatus driver, what one behavior can you change? As the company officer, what one behavior can you change? As a chief officer, what one behavior can you change? Good questions for your next drill. Your answer may help save a life . . . including yours!

References

Brunacini, A.V., 2008. "Fast /Close/Wet: Reducing Firefighter Deaths and Injuries: Changes in Concept, Policy and Practice." Public Entity Risk Institute, Fairfax, VA.

Kunadharaju, K., Smith, T. D. and DeJoy, D. M., 2011. "Line of Duty Deaths among U.S. Firefighters: An Analysis of Fatality Investigations. Accident Analysis and Prevention." 43, 1117-1180.

National Fallen Firefighters Foundation, 2004. "Everyone Goes Home: 16 Firefighter Life Safety Initiatives." NFFF, Emmitsburg, MD.

Schein, E.H., 2004. *Organizational Culture and Leadership.* Jossey Bass, San Francisco, CA.

Fire/EMS Safety & Health Week: Rules vs. DNA

2012

THE 2012 INTERNATIONAL Fire/EMS Safety & Health Week theme is "Rules you can live by." If you have been in the fire service for twelve months, you know the rules that will keep you from getting injured, killed, fired, voted out, or disciplined. You have also seen or read about most of these rules broken by other firefighters with no negative consequences. The longer you are in the service the more rules you will see broken or not enforced and sometimes even rewarded.

Not following the rules is part of our culture and what the public has come to expect from us. Society gives us permission not to follow the rules. We get these special privileges because people call us when they are having a really bad day that is getting worse and endangering their lives and property.

Our country was founded on the principle that people are endowed by their creator with inalienable rights, among which are life, liberty, and the pursuit of happiness. When your house is on fire, your life, liberty, and pursuit of happiness are in jeopardy. If nothing is done, your neighbors' rights are at risk as well. Left unchecked, the entire community can be lost.

From the beginning of the United States of America, society has expected firemen (today we are called firefighters) to come fix the problem. We have done this willingly with great pride, skill, courage, and sacrifice. All the while we are being cheered on by an adoring public. Even our death is commended by God: "Greater love has no man than this that a man lay down his life for his friends" (John 15:13). How many times have you heard this Bible passage at a firefighter funeral? When we get killed, it is a line-of-duty death, and the whole town turns out to show respect and share grief. We have been doing this for almost three hundred years.

The number-one Life Safety Initiative of the National Fallen Firefighters Foundation's "Everyone Goes Home" campaign is, "Define and advocate the need for a cultural change within the fire service relating to safety, incorporating leadership, management, supervision, accountability, and personal responsibility." When the leaders of the fire service wrote this more than eight years ago, I was there. They knew the firefighter injury-and-death problem was not a lack of rules, or lack of knowledge of the rules, or lack of training on the rules, or lack of the ability to follow the rules. It was the fact that firefighters and fire departments pick and choose what rules to follow and what rules to ignore. In 1974, renowned author Frank Brannigan said, "We are not killing firemen in any new ways." Brannigan's statement was confirmed by a 2010 study. (See page 40.)

What are the paradigms of firefighting that make up our culture of longstanding acceptance and tradition? What are the operational tenets that are accepted as the way we do thing? What is it that the public and fire service reinforce through socialization and positive public image? In other words, to understand culture we must identify the underlying assumptions that drive our behavior (Schein, 2004). What is it in firefighter DNA that drives our behavior and the public reinforces?

From Ben Franklin to today, all firefighters have the same DNA made up of six genes: fast, close, wet, risk, injury and death. These genes have been passed down for generations from firefighters and the public. Our gene sequence has driven our behavior and rule development throughout our history (Clark, 2011).

Fast. In the beginning, firemen ran to the fire as fast as they could, pulling the hose cart and carrying the buckets. Today we drive the apparatus as fast as we can, with permission to ignore public traffic laws because we must be fast. If we crash into someone on the way to the fire and kill them, we are shielded, in some states, from gross negligence because we must be fast. If a firefighter in the crash gets killed, it is an LODD and the family will receive a $320,000 federal death benefit. The benefit is paid even if the firefighter was not wearing a seatbelt. In some states, firefighters are not required to use seatbelts when responding because they must be fast. Finally, the worst thing that can happen is to get beaten to your first-due area by another company or have another fire company steal your fire. Even NFPA standards require us to be fast: We must be out of the station in sixty seconds.

Close. When you have only a bucket of water to throw on the fire, you have to get very close. Today we are trying to develop a higher-temperature facemask lens so we can get closer to the fire because the present plastic lens melts at 536 degrees Fahrenheit.

Wet. In my forty-two-year career, we have gone from 2½-inch and 1½-inch to 1¾-inch, two-inch, three-inch, and five-inch hoselines. We've used slippery water, Class A foam, High-X foam and compressed air foam and gone from 350 GPM to 2,000 GPM pumps. Yet, when you read the NIOSH reports of these deaths in building fires, in many cases water was missing either from an attack line or fire sprinkler. When the public and firefighters think wet, we see big red firetrucks, not little fire sprinkler heads.

Risk. It's what makes us special in our own eyes and the minds of the public. We, and the public, believe firefighters will save their life, property, and community. "Risk a lot to save a lot—risk nothing to save nothing" is our slogan today. But it is up to each firefighter and each fire department to define *a lot*, *nothing*, and *risk*. The slogan did not help in February 1973 when my community lost seven people in three house fires. They were dead before we left the station. This was before residential smoke alarms. Our risk slogan did not help in March 2012 when nine people died in a house fire that you could see from the fire station because there were no working smoke alarms. The fire service and public give the highest awards to the firefighters who take the greatest risks. Nobody gets awards for installing smoke alarms. Some states even outlaw residential fire-sprinkler codes. Courage and valor awards are given to firefighters who do not follow the rules, because fire warriors take risks.

Injury. It's the firefighter's red badge of courage and the public recognizes it. Your third-degree burns may be referenced when you are appointed as fire chief or you may get a congressional or manufacturer's award when you get out of the burn unit. Firefighters who are injured are rarely disciplined or denied worker's comp, even if they were not following the rules. After all, risk and injury go hand-in-hand, so firefighter injury is part of the job.

Death. "Firemen are going to get killed. When they join the department they face that fact. When a man becomes a fireman, his greatest act of bravery has been accomplished. What he does after that is all in the line of work. They were not thinking of getting killed when they went where death lurked. They went there to put the fire out, and got killed. Firemen do not regard themselves as heroes because they do what the business requires." This 1908 quote from FDNY

Chief Ed Croker, spoken upon the death of a deputy chief and four firemen, has been repeated throughout my time in the fire service, even by the vice president of the United States. In 1976, my lieutenant paraphrased the death gene when he stated death is part of the job. Even today, some of my contemporaries state, "Not Everyone Goes Home." Chief Alan Brunacini shines light on our "Rule vs. DNA" death gene abnormality. (See page 42.)

Your firefighter DNA genes (fast, close, wet, risk, injury, and death) will trump rules every time. Most of the time, one abnormal gene does not negatively affect the outcome, but when two or more mutate, tragedy can result. Changing your DNA is hard, but you can change your behavior if you know what is driving it.

In January 2012, a nineteen-year-old volunteer who had just completed Firefighter 1 was killed in a single-car crash responding to a garage fire in his personal vehicle at high speed with no seatbelt. Our FRD (fast, risk, death) genes killed him. In April 2012, a sixty-year-old career lieutenant and a twenty-five-year-old career firefighter died when a wall they were inspecting collapsed on them, twenty-nine minutes after the five-alarm fire in a vacant warehouse was brought under control. Our CRD (close, risk, death) genes killed them. I pray our three brothers rest in peace along with all the other LODDs, past and future.

Whether you are a firefighter for twelve months or forty-two years, you pick Rules vs. DNA to live by every day on every call. At your next drill, consider firefighter gene therapy for fast, close, wet, risk, injury, and death. It may let you go home one more time.

References
Brunacini A.V., 2008. "Fast/Close/Wet. Reducing Firefighter Deaths and Injuries: Changes in Concept, Policy and Practice." Public Entity Risk Institute, Fairfax, VA.
Clark B.A., 2011. "Your behavior comes from Ben Franklin's DNA"

Firehouse.com, Oct 2011. http://www.firehouse.com/topic/leadership-and-command/your-behavior-comes-ben-franklins-dna-fast-close-wet-risk-injury-death

Clark B.A. 1976. "I don't want my ears burned" Fire Command, July 1976, p.17

Kunadharaju K., Smith T.D. and DeJoy, D.M., 2011. "Line of Duty Deaths among U.S. Firefighters: An Analysis of Fatality Investigations." *Accident Analysis and Prevention*. 43, 1117-1180.

National Fallen Firefighters Foundation, 2004. "Everyone Goes Home: 16 Firefighter Life Safety Initiatives." NFFF, Emmitsburg, MD. Schein, E.H., 2004. Organizational Culture and Leadership. Jossey Bass, San Francisco, CA.

Why the Fire Service Should Avoid the Term 'Line-of-Duty Death'
2013

I HAVE NEVER BEEN in the military. My daughter Samantha served in the first Gulf War as a medic in a combat zone; her son Stryder just graduated Advanced Individual Training school at Fort Lee, Virginia, as a 92F specialist. Of course I was concerned for Samantha's safety when she was deployed, as I am concerned about my grandson's potential deployment—but none of my friends or relatives in the armed services was ever killed in action. Although I have an eternal gratitude to all military service members who gave the ultimate sacrifice to make and keep us free, my direct connection to the concept of being KIA is fairly limited.

I do, however, tear up when "Taps" is played, just as I cry at firefighter funerals, although I've been blessed never to be on an incident where a firefighter died.

Why do I bring up KIA in conjunction with firefighter funerals? Since 1970, when I joined the fire service, I have been uncomfortable with the comparison of firefighters to soldiers in combat. At a 2012 national lecture I made the comment that "comparing a firefighter death to a soldier's death is an insult to the soldier." To my knowledge, that's the only statement from this ninety-minute lecture that was shared on social media. Yet we continue to make the comparison.

The military and the fire service have different missions. In war, military personnel are trying to kill each other to defend their governments and societies or to claim additional power for their group or country. The fire service and firefighting are about saving lives. What we do is not war or combat.

Defining KIA

We all know what KIA means, right? Actually, the definition is a bit more complicated than you might have thought. According to Wikipedia:

> Killed in Action (KIA) is a casualty classification generally used by militaries to describe the deaths of their own forces at the hands of hostile forces. The United States Department of Defense, for example, says that those declared KIA need not have fired their weapons but have been killed due to hostile attack. KIAs do not come from incidents such as accidental vehicle crashes and other "non-hostile" events or terrorism. KIA can be applied both to front-line combat troops and to naval, air, and support troops. Someone who is killed in action during a particular event is denoted with a † dagger beside their name to signify their death in that event. Further, KIA denotes one to have been killed in action on the battlefield whereas died of wounds (or DOW) relates to someone who survived to reach a medical treatment facility.

The fire service equivalent of KIA is the line-of-duty death—yet in contrast to the military's narrow definition of KIA, the fire service classifies all firefighter occupational fatalities as LODDs. This includes a heart attack twenty-four hours after responding or training, dying at a training activity, falling from a firetruck that is moving or parked, dying in a vehicle crash when not wearing a seatbelt, dying from being struck by a moving vehicle, or dying on a treadmill while exercising at the fire station.

Using the term *LODD* perpetuates the myth that firefighters' getting killed is part of the job.[1] When we use these words, we are continuing the tradition started almost three hundred years ago in the era of Ben Franklin. Ben's firemen were volunteers trying to protect their neighbors from the ravages of fire. There was great risk to the firemen, and deaths did occur. When a fireman laid down his life to help his neighbor, the entire community felt the loss and commemorated their friend as a hero—regardless of the cause of death, such as slipping on the wet cobblestone street or being run over by the hose cart. (In Ben Franklin's days it was entirely possible for firemen to have just left the pub to pull the hose cart to the fire. Needless to say, they didn't do blood alcohol testing!)

After the Civil War, many military-trained men became the first paid firemen. Honoring fallen firefighters in the same manner as fallen soldiers was a natural extension of the military tradition into the fire service. Society and the fire service continue, to this day, to see almost any firefighter's death as the ultimate sacrifice for the community.[2]

Chief Alan Brunacini has noted that this tradition is problematic. This culture has been noted among researchers as well. Indeed, the fire service and the public rarely hold anyone accountable for a firefighter's death. Some state laws actually shield firefighters from gross negligence related to another firefighter or civilian death.[3]

Trying another term

What if we replaced the term *LODD* with the term *occupational fatality*? Again, Wikipedia provides some clarity into whether this is a fitting definition for the fire service:

> An occupational fatality is a death that occurs while a person is at work or performing work-related tasks. Occupational fatalities are also commonly called "occupational deaths" or "work-related deaths/fatalities" and can occur in any industry or occupation. Common causes of occupational fatalities include falls, machine-related incidents, motor vehicle accidents, electrocution, falling objects, homicides, and suicides. Occupational fatalities can be prevented.
>
> Many factors contribute to a fatal incident at work. Lack of appropriate employee training and failure to provide and enforce the use of safety equipment are frequent contributors to occupational fatalities. Incidents can also be the result of insufficient supervision of inexperienced employees or employees who have taken on a responsibility for which they are not properly trained. Poor worksite organization, staffing and scheduling issues, unworkable policies and practices, and workplace culture can all play a role in occupational fatalities. In any case, the incident leading to an occupational fatality is generally not the fault of a single person, but the tragic result of a combination of many human and environmental factors.
>
> Occupational fatalities are preventable. Prevention of occupational fatalities depends on the understanding that worker safety is not only the responsibility of the worker, but is the primary

responsibility of the employer. Employers must train all employees in the appropriate safety procedures and maintain a safe working environment so that fatalities are less likely to occur. An occupational fatality is not just the fault of the deceased worker; instead, it is the combination of unsafe work environments, insufficient safety training, and negligible employee supervision that contribute fatal incidents. As a result, it is imperative that an employer address all the potential risk factors at the workplace and educate all employees in safe work practices and risk awareness.

A key part of the definition: An occupational fatality is not part of the job—rather, it's an indication that something went wrong. Corrective action and accountability are applied so it does not happen again.

A different standard

Society does not hold the fire service to the same occupational safety standard as other industries. To illustrate this, let's compare the Deepwater Horizon tragedy to the Charleston Sofa Super Store fire.

Following the Deepwater Horizon incident, British Petroleum pleaded guilty to felony manslaughter for the deaths of eleven workers, as well as environmental crimes and obstruction of Congress. BP paid four billion dollars in fines and penalties.[4]

The city of Charleston, on the other hand, "does not admit the truth of any alleged facts, any of the characterizations of Respondent's alleged conduct, or any of the conclusions set forth in the citation issued in this matter." The South Carolina Office of Occupational Safety and Health "has made no representations or determinations concerning the probable cause of the injury or death of any person involved in the June

18, 2007, fire." The fine for the Sofa Super Store fire—where nine firefighters were killed—was $3,160, but not for the dead firefighters. The citation was for three or more firefighters at the incident who weren't wearing their personal protective equipment, three or more firefighters at the incident who weren't wearing SCBA (excluding the dead firefighters), and the fact that the "employer knew or should have known that its written incident command system does not directly address those emergency response situations which do not include a HAZMAT incident."[5]

I compare BP to Charleston not to find fault with any government agency, but to illustrate how the public and the fire service has a long tradition of acceptance of risk and loss of firefighters. This will continue if we perpetuate firefighter LODDs as just part of the job.

The power of words

"Line-of-duty death" is a twentieth-century term, just like "firemen." Firefighters are not all men, and when a firefighter is killed it's not part of the job. For the twenty-first century, the fire service and our society—the people we protect—need to understand firefighter causalities as occupational injuries and deaths that are preventable.

Ceremonies are important to the fire service and society. They communicate our values and beliefs about what we hold dear. If Chief Brunacini is correct, elaborate ceremonialization of battleground deaths should be reserved for soldiers killed in action and for firefighters who were taking an appropriate risk to protect a savable life. But even a change in our funeral ceremonies isn't enough. We must also conduct mandatory, comprehensive casualty investigations, including inquests into the responsibility for firefighter occupational deaths. When we put those report references under our lost brother or sister's picture hanging at the fire station, and every firefighter nationwide reads the report and looks into the face of our lost firefighter, we will not only remember them—we will learn the lesson they paid dearly to teach us.

Words are powerful. We need to change some of our words to help ensure "Everyone Goes Home." Let's start by changing how we refer to firefighter line-of-duty deaths.

References

1. Clark BA. (1976) "I don't want my ears burned." *Fire Command*, July 1976, p.17.
2. Clark BA. (Oct. 2011) "Your behavior comes from Ben Franklin's DNA," Firehouse.com. Retrieved 12/26/12 from www.firehouse.com/topic/leadership-and-command/your-behavior-comes-ben-franklins-dna-fast-close-wet-risk-injury-death
3. Clark BA. (June 2012) "Fire/EMS safety and health week: Rules vs. DNA." In Firehouse.com. Retrieved 12/26/12 from www.firehouse.com/article/10732003/fire-ems-safety-health-week-rules-vs-dna?print=true.
4. U.S. Department of Justice (Nov. 15, 2012) "BP Exploration and Production Inc. Agrees to Plead Guilty to Felony Manslaughter, Environmental Crimes and Obstruction of Congress Surrounding Deepwater Horizon Incident." In Justice.gov. Retrieved 11/19/12, from www.justice.gov/opa/pr/2012/November/12-ag-1369.html.
5. State of South Carolina. (2007) Office of Occupational safety and Health of the South Carolina Department of Labor, Licensing and Regulation vs. City of Charleston. Columbia: 961767 v. 7

Editorials Thirty-Seven Years Apart Agree: Getting Killed Is Not Part of the Job

2013

IN 2013, ASSISTANT professors at the University of Idaho and Harvard and the incoming president of the IAFC agree with my 1976 editorial that firefighters getting killed is not part of the job. (See page 21.)

Editorials are just opinions—and everyone has one—but some opinions are more important than others. That is what leadership is all about: When people in positions of power think or feel a certain way it can drive their behavior to influence others to cause change and outcomes. The IAFC, Harvard, and U of I are leadership organizations so when individuals representing them speak, society listens.

President Kennedy said we will put a man on the moon and return him safely; when he said that the technology to do it was not available. But his opinion, vision, leadership, dream, and position of power made it happen. Does the American Fire Culture need a JFK?

Bill Metcalf EFO CFO is the first vice president of the IAFC and fire chief of the North County Fire Protection District in Fallbrook, California. He will become IAFC president on August 16. An editorial recently written by Metcalf explains why he is angry at the loss of nineteen Granite Mountain Hotshots at the Yarnell Hill Fire:

> In spite of all we know about fire, in spite of all of the advances that have been made in technology, in spite of all of the advances in science of fire prevention and suppression, in spite of the billions of dollars spent—firefighters still die.... We mourn, we grieve, we study, we write reports, and we publish statistics, yet we continue to lose firefighters in fire —wildland and structural.... Is that the best that we can do? Is that all we've got? ... I believe we hold the power today to stop all firefighter line of duty deaths. Yes, all of them.... What we lack is the will to make it happen.... We'll have to challenge and put away many traditions that many of us hold dear. And we'll have to think in new, brave ways. We know exactly what must be done.

Dr. Matthew Desmond, an assistant professor of sociology and social studies at Harvard, served four seasons as a wildland firefighter in Arizona. His July 6 editorial in the *New York Times* explains how firefighters understand death and why that is dangerous:

> The writer Joan Wickersham has said that tragedies cause us to think two incompatible thoughts: *Why* and *Of course*. We search for answers while already knowing them.... For firefighters, there is no of course.... Those nineteen, many will think, must have made a mistake. They lacked a proper escape

route, some officials now surmise.... As one of my crew members told me in 2003 after a Hotshot supervisor died in a nearby forest, "You really have to do something wrong when you're dealing with fire, to be in a situation where it takes your life." ... One of my crew members liked to say, "I trust one person and that's myself."... But we cannot claim with perfect certainty that entrapments are always a result of violations of the ten and eighteen. Plus, some of the orders are fuzzy. What does it mean, exactly, to "Fight fire aggressively, having provided for safety first," the tenth order? ... Too much self-reliance among firefighters is dangerous.... According to several studies, the source of many fatal entrapments has been firefighters' can-do attitude or a culture emphasizing individual work rather the group work. This idea, which pivots upon the importance of the individual, not the group, might make firefighting more dangerous than it has to be.

Dr. Crystal A. Kolden, is an assistant professor of geography at the University of Idaho, where she heads the pyrogeography lab; she is a fire ecologist and former U.S. Forest Service firefighter. Her July 5 editorial in *The Washington Post* illustrates how society must take some responsibility for the nineteen firefighters killed in Yarnell:

> They should never have been put in that position.... Since Yarnell had already been evacuated, these men were lost trying to save not lives but houses.... Homeowners who live in wildfire-prone areas shouldn't expect their highly flammable property to be rescued during extreme fires.... For example, more than a million new homes were built

in high fire danger arrears in California, Oregon and Washington since 1990. But homes are built amid dense, flammable vegetation everywhere from Florida to Michigan to Texas to New Jersey.... We need to recognize that communities built without wildfire-mitigation measures are tinderboxes waiting to burn and stop incentivizing homeowners to rebuild with kindling.... The firefighting community will come up with a new set of safety standards and go through more training, and the public will mourn the dead and then move on to the next tragedy.... Communities where wildfires have consumed homes this summer will receive their insurance checks and rebuild. Things will go back to business as usual for everyone—except the loved ones of those lost.... This cycle must end.... Insurance companies must either charge more to insure highly flammable homes in wildfire-prone areas or abandon such areas entirely to reflect the true dangers homeowners face.... Fire agencies need to stop using outdated suppression tactics that are too dangerous for more intense fires that burn in more extreme weather amid more houses.... Wildfires are not preventable....Unless we fundamentally change the way we view wildfires, brave men and women will continue to put themselves in harm's way to protect our homes. And we will have to explain to the families and loved ones of those lost why we thought our houses were more important than their lives.

My 1976 editorial predicted future firefighter deaths and injuries. The 2013 editorials say that firefighter deaths can be stopped. Will the American fire culture repeat the seventy-five total LODDs so far in 2013? Who will be our JFK who says firefighter occupational fatalities will be stopped by 2018 and then make it happen? It should be easy because we already know how to do it, but it will take our collective courage to make this opinion a reality.

The Ultimate Ethics of the Badge
2014

FIRE SERVICE IS different from other disciplines because of the risk to individuals, society, and firefighters at work. We carry a badge and are held to a higher ethical standard than citizens or businesses. The National Society of Executive Fire Officers published a model firefighter code of ethics but it does not address the ultimate ethical question of life and death in the work of firefighting.

What is ethics? Synonymous to moral theory, ethics systematizes, defends, and recommends concepts of right and wrong conduct, and often addresses disputes of moral diversity. Ethics is divided into three major areas: 1.) meta ethics, the theoretical meaning, reference of moral propositions, and how truth values are determined; 2.) normative ethics, the practical means of determining a moral course of action; and 3.) applied ethics, which draws upon ethical theory to ask what a person is obligated to do in a specific situation or particular domain of action.

Let's examine the firefighter-death question at these three levels of ethics.

Meta-ethics. Since the beginning of time, humans have needed to protect themselves from nature and each other. Preservation of self and family is a human condition of existence. The protectors are held in high esteem, and the ultimate protection results in war.

Humans go to war and try to kill each other. The shield has been part of the warrior equipment for thousands of years. Some organizations refer to their badges as shields. The fire service uses the term "fire warrior"—a message that death is part of the firefighter's job.

When is it acceptable or ethical to society for a firefighter to die doing their job? Newspaper headlines on firefighter fatalities usually read "The City Lost Two Heroic Firefighters Today." The headline never says "The City Killed Two Firefighters Today for No Good Reason."

Please send me a copy if your job description or volunteer application form contains any form of this sentence: "As a firefighter, you are expected by society to risk your life doing your job, which can result in your death." Does this moral proposition exist, whether explicit or implied? This is the meta-ethical question that needs to be answered for the fire service in the twenty-first century if we truly want everyone to go home.

Normative ethics. When we put the badge on, does society expect us to die doing our job? When someone we love pins the badge on our chest, do they think getting killed is part of the job? When we get on the truck for the first or the ten-thousandth time, is there a possibility we would not return alive? Yes. But is our death ethical? Is there a right and wrong firefighter death?

If the fire turns out to be arson, society has a person to blame for the death of the firefighter, so the death is wrong and the firefighter is a hero. This remains true even if the firefighter, officers, and commanders violated every SOP and safety standard during the incident. The situation is similar if the building has a fire- or building-code violation, because the building owner can be blamed, regardless of how the fire department operated.

How many times has the justification of going into an abandoned building to search for a vagrant been used to explain a firefighter's death? If the firefighter believes someone is in a burning abandoned building, permission should be sought to go beyond our standards, to ignore our SOPs and safety guidelines, and if tragedy occurs the death of the firefighter is heroic.

Applied ethics. As a firefighter, driver, officer, or organization, you use applied ethics every day and on every alarm. Your basic ethical question is: "What am I obligated to do, and is my behavior right or wrong in meeting the obligation?"

Am I obligated to report for work on time? Am I obligated to be sober at work? Does the organization hold me accountable to the obligation? Am I obligated to wear my seatbelt in the apparatus? Am I obligated to stop at red lights when responding?

What are the consequences of wrong behavior? This will help you in decision-making. Can I be fired for wrong behavior? Can I kill someone else with my wrong behavior? Can my wrong behavior kill me?

The ultimate ethics of the badge at the meta, normative, and applied levels is about life and death. What does your code of ethics say about that? Remember: Ethics matter!

Firefighter Philosophy
In One Word: Why?

2014

PHILOSOPHERS DON'T THINK or write about firefighters, and very few firefighters think or write about philosophy. But that does not mean philosophy and firefighting are not connected at a very critical level, which accounts for the manifest identity or core mission of the fire service: fight fire, save lives, and save property. Wikipedia can supply us with a working definition of philosophy:

> Philosophy is the study of general and fundamental problems, such as those connected with reality, existence, knowledge, values, reason, mind, and language. Philosophy is distinguished from other ways of addressing such problems by its critical, generally systematic approach and its reliance on rational argument. In more casual speech, by extension, "philosophy" can refer to "the most basic beliefs, concepts, and attitudes of an individual or group…"

The word "philosophy" comes from the Ancient Greek which literally means "love of wisdom."

We see the power of philosophy to influence all the time. Society decided that drinking and driving were unacceptable thanks to Mothers Against Drunk Driving. Sexual assault by priests on children became unacceptable thanks to Survivors Network of those Abused by Priests and, most recently, thanks to a new pope.

These notions go to the basic philosophical questions about our fire culture in society, the discipline, organizations, or groups. There is a philosophical aspect to every human action, reaction, thought, feeling, or belief. Socrates was teaching us to examine the why behind each of these throughout our life.

Let the examination begin.

If you have sex at the fire station, chances are you will be fired, but if you drive the chief's car over one hundred miles per hour, no problem. These examples are not to disparage any individual fire department or firefighter. They are to illustrate how culture and the philosophy behind it help us identify what is acceptable and unacceptable behavior.

When we examine the cultural foundation to some specific behavior, we are trying to answer the philosophical question "Why?" Why did the firefighter not use his seatbelt? Why did the firefighter disable the seatbelt alarm? Why was he speeding? Why did he drive through the red light? Why did he die when ejected from the apparatus? Why do we consider it a line-of-duty death? Why does the federal government give his survivor $330,000 death benefit? Why do we treat his death as heroic? Why did the house catch fire? Why was the smoke alarm not working? Why does the state law forbid mandatory residential fire-sprinkler laws in new home construction? Why did a mom, dad, and their two children die in the house fire?

What is the philosophical wisdom that drives these next four examples of cultural artifacts and the underlying assumption they are based on? The death of nineteen firefighters in Arizona was investigated by two separate groups that came to very different conclusions. Fire service conclusion: Nothing went wrong. Occupational safety conclusion: willful, serious violations resulting in death. A state legislature and governor overturned a mandatory residential-sprinkler law for new residential occupancies. The same legislature and governor approved a cancer presumption law for firefighters.

My twenty-two-year-old grandson, Gage, just began his journey down the philosophical road to wisdom when he said, "Grandpa, when I was a kid I thought grownups had all the answers, and now that I'm older I realize they don't know all the answers." My reply, as a sixty-four-year-old, was, "Yes, we are all trying to figure life out."

Whether you are the newest firefighter on your crew or the senior officer at the incident, how well you examined life before you arrived may determine the outcome. Unfortunately, philosophy is not currently included in any fire-service curriculum, from basic to the most advanced. Luckily, asking why is part of our basic human nature and it may be what makes us human.

The American Fire Culture video helps explain why: It is hard to ask the philosophical question: Why do we have this culture? By asking the hard questions about culture, we help ensure our survival as a species at the human level and help ensure everyone goes home at the firefighter level.

If you want to begin a concise philosophical journey to fire-service wisdom, read the references and ask yourself why about the content, the author, and most importantly your reaction to the reading?

Fire Service Philosophy 101 begins. At the personal level, it may help you understand why you behave the way you do on your next call and go home after it. At the societal level, it may help you understand our fire culture and the why behind it. Understanding why is always the first step to wisdom, and just by asking the question we are living an examined life.

Seatbelts Save Lives:
Why Don't We Wear Them?

To Be or Not To Be a Tattletale
2003

SHAKESPEARE AND *TATTLETALE* are not usually combined in a title but these two concepts represent the dilemma I faced. In the first concept, "To be or not to be," Hamlet is contemplating whether it is better to be alive, facing all the trials and tribulation of life, or to be dead and not have any responsibilities. Second, we learn the concept of "tattletale" at a very young age. I would get into trouble if I ran to Mom or Dad and told on my younger brother. This was a useful childhood rule because my brother was not allowed to tell on me. As adult fire-service professionals who make life-and-death decisions, we cannot let childhood concepts misguide our responsibility to each other.

I had just completed teaching a class on the constructs of vision, mission, and values to a group of fire officers. My plane did not leave for several hours, so the local fire station invited me to dinner. The menu included Sloppy Joes and homemade coleslaw, excellent. I asked if I could ride out on calls, no problem. The crew was proud of its new rescue pumper and eager to show it off.

We get a call for a stuck elevator. We were the fourth due unit. We respond lights and siren. The new rescue pumper has three-point seatbelts that are bright orange so it is easy for the officer to see that the safety devices are being used. Four of us respond; only one seatbelt is used. One firefighter stands up while the vehicle is moving to put his bunker pants on. After the call, only one seatbelt is used on the drive back to the station.

The fire chief picks me up after dinner for the ride to the airport. Do I tell him that the seatbelt policy is not being enforced? Does he already know? Will he think I am a tattle-tale? The crew that treated me to dinner are great guys. I'm not a squealer. I do not have any official authority or responsibly to address this issue. The twenty-minute trip to the airport is long. I say nothing.

The plane ride home was even longer. What if the crew has an accident on its next run and one of them is injured or killed—how would I feel? If we'd had beer with dinner would I tell? Would I have gotten on the apparatus? I make my crew wear their seatbelts; I am not very popular.

What do we hold each other responsible for? What am I responsible for? When one of us is injured or killed because of not following the rules, did anyone know, did anyone care, did anyone try to correct the behavior before it lead to a tragedy? Am I my brothers' keeper?

From now on, if I am on your firetruck and the seatbelts are not used I am going to ask you why. Then I am going to tell the chief that the seatbelt policy is being ignored. Then I am going to publish the name of the fire department, the chief, and the company officer. I got that idea from Mothers Against Drunk Drivers. You have all been warned. It is your choice: Wear the seatbelt or do not invite me to dinner and do not let me ride along.

I have decided that professionals, career or volunteer, do not look the other way. We pride ourselves on our willingness to risk our life to save each other heroically; yet we do not have the courage to be responsible for each other's safety behavior. This contradiction must change.

I will lose friends and miss dinners because I never want to say, "I could have saved a life that day, but I chose to look the other way." (See page 34.) Do you look the other way?

How to Get Firefighters to Wear Seatbelts
2004

ONLY FIFTY-FIVE PERCENT of firefighters wear their seatbelts.[1] The number-two cause of firefighter line-of-duty deaths is vehicle crashes. There is a direct correlation between these two facts. But the bigger tragedy is that we, the fire service, know this problem exists and have not fixed it. If you are a chief, company officer, or a firefighter, you are responsible for fixing this problem in your department. There is no excuse for not wearing your seatbelt and no excuse for not enforcing the fire department's seatbelt rule.

How do you get firefighters to wear seatbelts? You ask them to do it. This is an easy answer, but it is not a simple task. There are four steps to seatbelt-use success.

First, it takes soul searching. Are you and your department committed to firefighter safety, or is it just a slogan? Does your behavior reflect your true values? Successful use of seatbelts requires a two hundred percent proficiency score. This means that one hundred percent of the firefighters must wear their seatbelts one hundred percent of the time.

Second, it takes unlearning and relearning. Do your

firefighters race out the door, putting their bunker gear on as the apparatus speeds down the street because they do not want to get beat in by another company? Do your firefighters complain, "I can't get my SCBA on with the seatbelt." "People will die if I'm not ready to save them the instant the firetruck arrives at the fire." This is the same firefighter who doesn't wear the seatbelt on the return trip to the fire station. Someone put this dysfunctional thinking in his or her brain. You can learn to put your seatbelt on without increasing the fire loss in your community.

Third, it takes engineering. If your apparatus is so old that it did not come with seatbelts, have them installed. If you have only lap belts, have three-point belts installed. If for some reason the belts are not long enough to fit around your members, have longer belts installed. If there is a seat position without a seatbelt, do not let anyone ride in it.

Fourth, and most important, it takes leadership. Do you have the courage to do the right thing even when it will make you unpopular? Can you be the one person who stands up to the crowd because it is the right thing to do?

Can one person make a difference? Can your department serve as a role model for seatbelt usage? YES! The following case histories tell the seatbelt story of four fire departments and the leaders who had the courage to do what is right. I asked these chiefs to tell their seatbelt stories in their own words. We can all learn from them.

Washington, Ohio, Township Fire Department
By Chief Al Woo

One would think that communicating the need for vehicle safety would be relatively easy, especially for firefighters. It has been our experience, however, that this task was much easier to talk about than do.

Our seatbelt experience began with a simple observation made by another professional "riding out" as a guest on one of our vehicles. Dr. Burt Clark observed that many, if not all, of the firefighters assigned to that particular vehicle did not wear seatbelts to or from incidents.[2] I know it caused a great deal of concern to Dr. Clark, both in what he observed as well as to how to best inform me, the chief, about what he had experienced. After a couple of email exchanges, it became obvious that Dr. Clark's primary intentions were to ensure, as best possible, the safety of the firefighters. A second and more daunting realization was that since these issues were raised, I had an absolute obligation as chief to address them and keep our personnel safe.

On a subsequent visit, Dr. Clark, riding with another local department, had the opportunity to once again experience an "unbelted" response. This second incident eventually led to a very spirited discussion during a seminar attended by graduates of Ohio's Executive Fire Officer Program.

Addressing this issue became more than "just do it," as it became evident that safety is used often as a buzzword and is addressed when convenient. It also became obvious that addressing the seatbelt issues was going to be anything less than convenient. It also brought up other issues that were just not expected.

As an example, our department has always stressed a rapid response that included many components such as notification, multiple-station placements, and the development of apparatus that are safe and practical. Over the years, additions and practices were developed for which the vehicles were just not designed. One of these "updates" was the addition of forty-five walk-away SCBA brackets. Communicating that these brackets were not designed into the original apparatus and that their use was only an option if they could be safely integrated with seatbelts became quite a challenge.

A second unexpected issue was that we had firefighters who, because of their size, had difficulty using seatbelts, especially when wearing complete personal protective equipment.

Unfortunately, as with many fire departments, these personnel offer multiple safety challenges: One is the use of seatbelts, and the second is the firefighter's lack of physical fitness.

We as a department chose to address the issues in a number of ways. First, seatbelt use became an immediate topic on a couple of our promotional exams. It was a way of gauging whether the potential officers really had the best interests of their crews in mind and how they would address the issues that arose from seatbelt use. A second tactic was to immediately review the existing standard operating policy on motor vehicle safety to educate, train, and address issues that hindered the use of seatbelts. A third approach has been to work closely with the manufacturer of our apparatus (located locally) in addressing the issues that meet our safety concerns as well as the manufacturer's liability issues.

I am sure that many see their department's personnel, equipment, and practices in the experiences that I've described. While extremely proud of both the department and personnel, I am not naïve enough to believe that we are unique or perfect. In our zeal to ensure the safety of others, we often forget or forgo our own safety—something that is neither wise nor excusable. In the end, it has become quite a journey but, ultimately, if all else failed, we had to communicate to all involved that safety was everybody's responsibility and, yes, we had to "just do it."

Atlantic City, New Jersey, Fire Department
By Battalion Chief Bob Palamaro

Decreasing line-of-duty deaths and changing safety behavior are very reachable goals. The recipe that has brought some positive results in Atlantic City includes a passionate change agent, honesty to admit your faults, leadership to rally the troops, identification of key people in the organization to lead change, and persistence to reach the goal.

On December 7, 2003, I arrived at the National Fire

Academy to begin a two-week course entitled "Leading Community Risk Reduction." I was attending my second year of the Executive Officer curriculum, a program designed to develop future leaders in the fire service. The first day of the program always starts in the campus auditorium with an orientation program. Dr. Burt Clark was leading the orientation session that morning. He was lecturing on the importance of getting firefighters to wear their seatbelts.

One statistic that really hit home and aroused my passion that morning was the number of firefighters (one hundred) killed in the line of duty each year nationwide. Dr. Clark pointed out that heart attack while performing duties was the number-one killer of firefighters; vehicle accidents while responding to and returning from incidents was the number-two killer. Firefighters' not wearing their seatbelts was a significant contributing factor to the vehicle-response statistic. The session ended that morning with Dr. Clark personally challenging everyone in the audience to go back home and make an effort to impact the line-of-duty death statistic by mandating firefighters in their departments wear their seatbelts.

As fate would have it, Dr. Clark's wife, Carolyn, was one of the instructors. During the next two weeks, I had the opportunity to converse with Dr. Clark a few times. During our conversations, I found out that both Clarks like to come to my hometown (Atlantic City) to visit. Their next trip would be in January 2004. During that visit, Dr. Clark would spend some time riding in my battalion chief car during my tour of duty.

While I was thrilled to have Dr. Clark as a new friend, I spent the next couple of weeks after leaving the NFA lamenting his next visit to Atlantic City. I knew one of the things Dr. Clark would be paying attention to was the seatbelt behavior of my firefighters and, even more significantly, my own seatbelt behavior. Even though New Jersey state law and my department mandate seatbelt usage, the fact of the matter was that most people in the department chose to ignore both the law and the department order.

Pondering my dilemma, I realized if I was going to influence behavioral change to get seatbelt compliance in my department, I had to be honest with myself in admitting my own noncompliance. I had to address and change my own behavior first. I wore my portable radio strap over my left shoulder with the radio hanging down my right side. When I got into the chief's car, the radio rested on the seatbelt connection, making it difficult to buckle up. It was easier not to buckle up so that I could free my mind to concentrate on the more important task of leading personnel at the scene.

A quick study of my response patterns and behaviors before entering the chief's car allowed me to make a small adjustment to solve my buckle-up problem. Taking the radio off my shoulder before entering the car and laying it on the passenger seat allowed a smooth transition into the car, and buckling the belt became a simple task. I will admit that in the beginning I had to consciously think of the steps to follow, but soon thereafter the conscious effort became my new normal response routine.

The next hurdle was to convince my members to wear their seatbelts. It had been two weeks since coming home from the NFA, and I felt my seatbelt behavior had become consistent enough to challenge my people to wear their seatbelts. The last thing I wanted was to pose a challenge to them and have inconsistent behavior on my part lead to accusations of hypocrisy and a short-circuiting of the behavioral change goal.

I thought a good time to challenge the firefighters to change their seatbelt behavior would be with the first tour on New Year's Day: the challenge could be in the form of a New Year's resolution. New Year's Day morning, I went around to the three stations in my battalion to collect the daily paperwork. While at the stations, I asked the captains to assemble their members. I talked to them about the lessons I had learned regarding seatbelts on my visit to the NFA. I reminded them about the state law and the department order mandating seatbelt usage. I asked them to take the time to read the article Dr. Clark had recently written on firefighter seatbelt usage.[3] I challenged them to make a New Year's resolution to always wear their seatbelts while riding on the apparatus.

Changing behavior is difficult, and the success or failure of this initiative ultimately rested with each individual's personal commitment to change. Not to be overlooked is the role informal leaders can play in behavioral change. I identified a couple of well-respected, knowledgeable, and competent firefighters in my battalion. Getting their buy-in would go a long way in convincing the others to respond.

I explained to the company officers that it is a physical impossibility for a battalion chief to monitor every company responding to and returning from incidents. Next to the individual firefighter, the company officer had the most control over seatbelt compliance. If the members did not allow their rig to roll out of the station unless every member was buckled up, the problem would be solved. I also asked the captains to evaluate their apparatus for environmental factors that may hinder seatbelt usage.

We found that seven of our ten front-line apparatus had engineering issues that prohibited seatbelt usage or made it terribly uncomfortable when more than four firefighters were in the apparatus. A new recruit class had just graduated; this gave us a short-lived luxury of having five-member companies for the past few months. One engine and one truck needed immediate retrofit, as two of our larger-framed captains could not fasten their belts while wearing full turnout gear. A new, longer belt corrected the problem in the engine; relocating an SCBA and ordering a new seatback solved the ladder issue.

We are currently contemplating cab modifications on six of our engines to accommodate five people safely buckled up.

An EMS cabinet will be removed to allow the relocation of the interior seating; cargo netting and bungee cords will safely secure the equipment that was previously stored in the EMS compartment.

It has now been eight months since my return from the NFA. We have made significant strides in changing our behavior and re-engineering our apparatus for safety, but the job is not done. We must be persistent to stay the course. We still have our lapses and sometimes forget to buckle up, but it is happening with less frequency. We are still struggling with budget constraints to complete our cab modifications, and we must continue on that front.

Can you reduce line-of-duty deaths by changing your safety behavior and have everyone buckle up? Yes! It will take passion, honesty, leadership, and persistence.

Violet Township, Ohio, Fire Department
By Assistant Chief John Eisel

The use of seatbelts has been a topic of discussion in our organization for several years. It remains a topic of discussion at our monthly staff meetings. The issue was addressed by Chief Kenn Taylor and referred to International Association of Fire Fighters Local 3558 for review. With data such as firefighter death and injury reports showing why this is such an important safety issue, a taskforce of two company officers and two firefighters began its journey.

When the work of the taskforce was concluded, each shift was given a PowerPoint presentation of its findings and recommendations, which included the following:

- Mandate an emergency driving class for all personnel.
- Recommend CDL endorsements for all personnel.
- Review dispatch criteria: Do we need all apparatus to respond on an emergency for certain types of calls?
- Wear seatbelts at all times while vehicles are in motion.
- The officer and driver are responsible for making sure

all personnel are belted before the apparatus leaves the ramp.

- Train on using (wearing) seatbelts while in bunker gear and SCBA.
- Attach seatbelt extenders in apparatus as needed.
- Come to a complete stop at red lights, four-way stops, and negative right-of-ways.
- Proceed with "due regard" through controlled/regulated intersections after ensuring all other vehicles have stopped.
- Pay attention to posted speed limits and road conditions.

At this time, we have implemented the taskforce's recommendations other than the CDL endorsements. Several important factors have helped us to successfully implement these recommendations. During the "training" phase, a motorist struck a neighboring department's engine head-on while it was returning from a run.

Our department responded to that incident, and the first-hand witnessing of the damage left an eternal impression on those who witnessed it. Fortunately, the firefighters' injuries were not career-ending, but they were significant. If they had not been wearing their seatbelts, we could not begin to speculate how the outcome might have been different.

The second factor is that the recommendations implemented came from a taskforce of people committed to improving our personnel's safety. In the preliminary stages, we had discussed removing the SCBAs from the cabs to emphasize the importance of using seatbelts. Taylor's statement, "It is more important to wear our seatbelts than to put on our SCBAs" was controversial, but it motivated our members to find a viable solution to this issue.

As mentioned, this issue is still the first topic presented at our monthly staff meetings, as a reminder to our officers to keep this issue in front of the members. Compliance has been very good, but it did not happen overnight.

Burton A. Clark

Revere, Massachusetts, Fire Department
By Deputy Chief John Moschella

This study involved approximately twenty firefighters from my group. Since the culture in the department was generally to refrain from using seatbelts, my initial intent was to find out why fire personnel did not buckle up when riding in apparatus. A brief questionnaire, with the following questions, was distributed to the group:

- Do you use seatbelts when riding on apparatus?
- If your reply is *sometimes* or *no*, then why?
- Do you know if there is a seatbelt policy in the Revere Fire Department?
- How do you feel about a seatbelt policy?

Several weeks previous to this survey, company commanders were given Dr. Burton Clark's article "How Do You Get Firefighters to Wear Their Seatbelts?"[4] The intent here was to enlighten officers and provide them with background information. The theoretical objective was to put in motion a change model based around facilitative, informational, attitudinal, political, or, in this case, authoritative factors.

In brief, the survey showed that most firefighters did not wear their seatbelts. The most prevailing reason was that since they were never instructed to do so, they chose not to wear one. Few knew it was a departmental rule.

The next step was to moderate discussion pertaining to the seatbelt issue. An event that occurred in the department around this time facilitated this project.

By coincidence, the fire department had just sworn in fourteen new firefighters, who were immediately instructed that wearing seatbelts was mandatory according to department rules and regulations. The onus of responsibility was now on each company commander to enforce this regulation for the recruits. Consequently, as one might say, "What is good for the goose is good for the gander." Armed with the information from the article and empowered with the regulation, henceforth to be enforced, officers from each apparatus saw to it that all personnel buckled up.

The seatbelt issue is by no means legitimized. It will only become so if each officer and each firefighter makes a conscious effort to use the device. As the group commander, I shall continuously remind personnel. Hopefully, each firefighter will do the same.

National help for local campaigns

If you plan to become a local fire department seatbelt leader, you have backup from national organizations, private industry, and the law.

Fire-service organizations are focusing attention on seatbelts. The National Fallen Firefighters Foundation Line-of-Duty Death Prevention campaign has identified seatbelts as one of the sixteen initiatives of the program. At the 2004 International Association of Fire Chiefs annual conference banquet, the outgoing president, Chief Ernest Mitchell, and the incoming president, Chief Bob Dipoli, reminded all fire chiefs that it is their responsibility to ensure seatbelt compliance in their fire departments.

National Fire Protection Association 1500, Fire Department Occupational Safety and Health Program, states:

- Section 6.3.1: "All persons riding in fire apparatus shall be seated and belted securely by seatbelts."
- Section 6.3.2: "Seatbelts shall not be released or loosened for any purpose while the vehicle is in motion, including the donning of respiratory protection equipment or protective clothing."

The Fire Apparatus Manufacturers Association, the Congressional Fire Services Institute, the Fire and Emergency Manufacturers and Services Association, and the IAFC have endorsed NFPA 1901 Annex D, which outlines what needs to be done to retrofit fifty percent of the apparatus in service today. Item D.3, "Upgrading or Refurbishing Fire Apparatus," Recommendation 13, states: "Seatbelts are available for every seat and are new or in serviceable condition."

The United State Fire Administration's webpage, "Applied Research & Technology: Emergency Vehicle Safety" contains useful information and links to help with your seatbelt campaigns.

Fire service private industry is also trying to help us use seatbelts. The VFIS program "Operation Safe Arrival" focuses on adopting-safe driving practices for apparatus at intersections and seatbelt use by all firefighters. In addition, the VFIS accident and sickness policy pays an additional death benefit for a member who was wearing a properly fastened seatbelt at the time of the fatal motor vehicle accident.

FAMA has encouraged its members to promote seatbelt safety in their advertisements and company literature. It has suggested the following taglines for print media: 'The Best Firefighters WEAR SEATBELTS," "Wear Seatbelts and Live," "Buckle Up and Live," and "Seatbelts: You'll Live with Them."

Finally, in thirteen states it is the law that all firefighters wear seatbelts. In thirty-two states, seatbelt use is required only for passengers in the front seats.

Five states exempt firefighters from using seatbelts, and one state has no adult seatbelt law.[5] Do you know in which category your state fits? If your state has one of the less-stringent seatbelt laws for firefighters, the state chiefs association and state firefighters association clearly have one of their next legislation agenda items identified for them.

International advice

When I explained that forty-five percent of the members of the U.S. fire service were not wearing their seatbelts to Don Henry, CFPS, CD instructor, Auto/Diesel & Fire Apparatus Maintenance, Lakeland College, Vermilion, Alberta, Canada, at the recent IAFC conference, he solved the problem in one sentence, "They need a spanking!" Not wearing your seatbelt is childish. The safety of the fire service and firefighters is not child's play.

How do you get firefighters to wear seatbelts? You tell them to do it: "Put on your seatbelt." That's an order we can all live with because firefighters need it, spouses expect it, and families deserve it.

Author's note

It has been my honor to play a small part in these four seatbelt success stories. Yet I have been a failure in my own organization. Try as I may, the Laurel Volunteer Fire Department, Company 10, Prince George's County, Maryland, does not have a two hundred percent passing score on using seatbelts. The LVFD is an excellent organization with wonderful members and a proud history spanning 102 years of "Service for Others." We would all risk our lives to save each other heroically. I call on the members and officers of the LVFD who do use seatbelts to have the courage to help our brothers and sisters who do not to "See the Light and Buckle Up."

References

1. Fire Pole Question, Firehouse.com.. Nov. 10, 2003.
2. Clark, B.A.,"To be or not to be a tattletale," Firehouse.com, Oct. 1, 2003.
3. Clark, B.A., "199% correct is not a passing score in the fire service," Firehouse.com, Dec. 22, 2003.
4. Clark, B.A., "How do you get firefighters to wear their seatbelts?" Firehouse.com, May 14, 2004.
5. Grafton, B; R. Thompson, T. Gies, C. Dusil, C. Smith, B. Dowers, "Inter-Personal Dynamics in Fire Service Organizations," National Fire Academy, group course project "Seatbelt Usage in the Fire Service" (Seatbelt law analysis based on Oct. 2003 data), July 26, 2004.

Seatbelts: The Hugh Lee Newell Story
2005

"I DON'T NEED TO wear my seatbelt because no firefighter I know has ever been killed because they didn't wear their seatbelt." This is the latest excuse I was given by a firefighter for not using his seatbelt. He made the statement in front of the department's safety officer and other firefighters at a social event. Adult beverages were being served, so his illogical thinking may have been due to a loss of brain cells. The safety officer commented to me later that the department still has a lot of work to do to achieve a one hundred percent seatbelt-use-compliance rate.

The logic (excuse) for not following a safety procedure because nothing bad has ever happened to you or anyone you know goes beyond seatbelts; it is the root cause of our poor safety culture. We know the safety doctrine; we have the skill to perform; we have the equipment; but we choose not to make safety a priority. The fire service relies a lot on experience as the best teacher. If that experience includes not following safety procedures with no negative outcome, we perpetuate the wrong behavior. If the company officer reinforces the wrong behavior by not correcting it, doing it themselves, or not disciplining the wrong behavior, it is repeated. If the battalion chief sees the failure to follow safety procedures and turns a blind eye, the wrong behavior is accepted. If the fire chief knows the safety rules are not being followed, he or she is condoning a poor safety culture as the department standard.

The firefighter's excuse for not using a seatbelt is bad enough, but the logic I was given by a deputy fire chief from a large metropolitan fire department disturbs me even more. This chief officer is a national speaker and author whom I personally respect very highly. When I asked him what his fire department was doing to get seatbelt-use compliance his answer was, "Seatbelts are not a priority for us. You have to pick what is important." This justification for not enforcing seatbelt rules is not illogical; it is purposeful, thought out, intentional, and very dangerous. If he is correct, we need to eliminate our seatbelt SOPs and remove seatbelts from fire apparatus. I did not get to ask him if he used his seatbelt in the chief's car.

The fact is safety standards, SOPs, and equipment of today did come about because some firefighter—in many cases more than one—was injured or killed. The fallen firefighter's friends and fire department demanded that changes in equipment, training, standards, and operations be made so it did not happen again. We do not want firefighters to die in vain.

All aspects of firefighter safety must be a priority. If we follow only the safety doctrine that is convenient, our poor safety record will continue. Worse yet, if we only follow the safety rules after a firefighter we know is killed or hurt, what does that say about who we are and the price, in death and injury, we are willing to pay?

The problem is we do not know our history. So we are destined to repeat the mistakes of the past. At a deeper level we dishonor those who came before us and in many cases gave their lives so we can be safer today.

But one person can learn from the past and become a leader in creating a positive safety culture in the fire department. Engineer Duane Hughes, Engine 1 Columbus Mississippi Fire Department, is making a big difference in his department after he met Engineer Hugh Lee Newell. We can all learn for the following story.

Taking a stand on seatbelt use: Hugh Lee's story
By Duane Hughes

Traditionally, leadership in the fire service has been seen as reserved for the higher ranks. Rarely have foot soldiers established fire department policy or vision. Recently I was presented with an opportunity to change this standard. Simply stated, I challenged firefighters to use seatbelts. Holding the rank of engineer, I was able to persuade many in my department that seatbelt use is not optional. With determination and a little courage, I proved leadership can spring from the lower ranks.

Two years ago I attended an Interpersonal Dynamics Course at the National Fire Academy. Dr. Burt Clark appeared in class and gave a speech concerning seatbelts and their lack of use in the fire service. Although I was not a supporter of seatbelt use, the forcefulness of Dr. Clark's speech struck a chord within me. When I returned to my department, I described the class to my station crew. After detailing the wonderful experience of the preceding two weeks, I mentioned Dr. Clark's speech. I remember telling the guys how I thought Dr. Clark was fighting a lost cause. "Not a lost cause, a just cause," responded Battalion Chief Truman Oswalt. Chief Oswalt was a long-time member of our department, and was affectionately known as "Hobby" by the guys. Hobby directed me to the hallway of our number-one station. Arranged along the walls were pictures detailing the exploits of our department. (Some of the older pictures date back to the late 1800s.) Hobby pointed towards an old black-and-white photo.

The framed picture shows a firefighter in an old-style dress uniform. Fastened to the bottom of the frame is a small metal tag which read, "Hugh Lee Newell, Sept. 11, 1931 / Oct. 1, 1972, Our Friend." Hobby fixed me with a stare and said, "I think you need to hear Hugh's story."

Hugh Lee Newell was a driver with the Columbus Fire Department. The apparatus was of the open-cab style and had no seatbelts. The captain and driver sat up front while the firefighter stood on the tailboard. In October 1972, Hugh and his crew were responding to an emergency call. While making their way through traffic, disaster struck. Swerving to avoid another vehicle, the front wheels of their apparatus struck the street curb. The firefighter was thrown from the tailboard, and narrowly missed being run over by the rear wheels. The captain maintained his seat, but Hugh was not as fortunate. Thrown from his position behind the steering wheel, Hugh landed in the truck's path. Unable to avoid his own vehicle, he was run over and killed.

While devastating, Hugh's death moved all the firefighters to action. Firefighter safety became the rallying cry of all who experienced the pain of Hugh's passing. Their impassioned pleas resulted in the retrofit of cabs to all Columbus Fire Department vehicles. This victory fell short of including seatbelts. The battle for seatbelts continued to rage on until 1984, when the retrofit of seatbelts was approved. Even this victory came with its own set of problems. Because of liability issues, the city garage and other local repair shops refused to install the seatbelts. Having come so far, the men refused to surrender the fight. Training Officer Kenneth Moore installed the first few seatbelts himself. Wearing full turnouts and seatbelts became standard procedure whenever an apparatus left the station. It was through these actions that the firefighters gave meaning to Hugh's death. The men of the Columbus Fire Department pledged themselves to safety, and strove to never again lose another friend to a preventable death.

After hearing the story of Hugh Lee Newell and the department's struggle for safety, I felt ashamed. How had attitudes in my department strayed so far from the ideals of 1984? Seatbelt use was no longer a battle cry, just a tired safety message. I believed that the lack of seatbelt usage in my department was an insult to the memory of Hugh Lee Newell. How many times as a firefighter had I refused to buckle up, believing it slowed my response time? How many times as a driver had I pulled away from the station, knowing that my passengers were not secured by seatbelts? I began to demand that passengers on my truck fasten their seatbelts. I was often met with resistance, but after hearing the story of Hugh Lee Newell, most firefighters agreed to fasten their seatbelts. Many other drivers began to take a firm stance on seatbelt use. When confronted with an unbelted captain, Driver Mike Chandler refused to proceed on a call. Later, Mike told me he was prepared to face dire consequences, but that truck wasn't moving until all seatbelts were fastened. Convincing stubborn firefighters to wear seatbelts is no easy task. My arguments for seatbelt use often fell on deaf ears. Many department members resisted change and saw the story of Hugh Lee Newell as ancient history. Several firefighters weren't born until well after Hugh's death in 1972. These younger firefighters simply couldn't relate to Hugh's story. That all changed with a visit from Mrs. Deana Vernon.

An opportunity for change came one station maintenance day. I washed the trucks as younger firefighters cleaned the downstairs quarters. Mrs. Vernon entered the station with her young daughter following closely. She remarked that the child loved firetrucks, and asked about the possibility of a tour. Presented with the opportunity to leave our chores and entertain the excited child, we happily agreed. After viewing the trucks and turnout gear, the tour proceeded inside the station. "Do you know the man in this picture?" asked Mrs. Vernon, while pointing to Hugh Lee Newell. "Yes, ma'am, he was one of our firefighters killed a long time ago," a young firefighter responded. Mrs. Vernon hugged her daughter and said, "Hugh was my father, and I am so touched that you guys remember him. I'm glad his death had some meaning. Just knowing all you guys can now wear seatbelts makes me happy." With tear-filled eyes, Mrs. Vernon recounted the media coverage of the department's fight for seatbelts. "It was always front-page news. I couldn't believe it took so long to get the seatbelts," she said.

What a victory! Mrs. Vernon accomplished in five minutes what I failed to do with weeks of reasoning. She put a face on her family's tragedy, and ended resistance to seatbelt use for all those young firefighters. Hugh Lee Newell would be honored by a new generation of seatbelt wearing firefighters.

I wish I could say that seatbelt usage was one hundred percent in the Columbus Fire Department, but that wouldn't be the truth. I know that cautionary tales and regulations won't change years of ingrained behavior. What I can say is that a change was made in my life after hearing the story of Hugh Lee Newell. My seatbelt is fastened every time I climb into the driver's seat, and my truck doesn't move until every passenger has seatbelt secured. I know that with each retelling of the Hugh Lee Newell story, another Columbus firefighter decides to buckle up. Leadership can spring from the lower ranks of the fire service. The fire service regularly displays courage and determination when dealing with public emergencies. Do we have the strength to display these same attributes toward our fellow firefighters? Can we love another firefighter enough to say, "Buckle Up"?

There are more apparatus drivers in the fire service than chiefs. When all drivers make seatbelt use a priority, only then can chiefs take seatbelts off their priority list because the department will be in compliance.

I want to thank Duane Hughes for sharing his story and demonstrating what leadership in the fire service is all about.

Battalion Chief Truman Oswalt deserves recognition for honoring Hugh Lee and Driver Mike Chandler deserves a courage award.

Finally, all firefighters, officers, and chiefs need to promise Mrs. Deana Vernon and our own family that we will wear our seatbelts. Because every life matters, even if you do not know them.

We Killed Firefighter Brian Hunton
2005

IF YOU DO not wear your seatbelt when riding on the firetruck, if you do not make your partner put his or her seatbelt on, if you drive the firetruck and all passengers are not buckled up, if you are the officer and you do not enforce the seatbelt policy, if you are a chief officer and do not hold your company officers accountable, if you are the fire chief and you know that you do not have a one hundred percent compliance one hundred percent of the time with your seatbelt policy—you killed Firefighter Brian Hunton.

I can hear the feedback to this charge now. "Clark, you have lost your mind! How dare you accuse me of killing a firefighter! Who the

> Firefighter Brian Hunton, twenty-seven-years old, with the Amarillo, Texas, Fire Department for two years, fell out of the ladder truck Saturday April 23 responding to a fire. Brian was putting his gear on, the door opened on a turn, and he fell out. He died from his injuries on Monday. My deepest sympathy to his family, friends, crew, and department. I hope we honor Brian by learning from this tragedy and change our seatbelt culture.

hell do you think you are? I don't know Brian. I'm not on his company. I wasn't driving the apparatus. I'm not his officer or chief. Firefighters have to take responsibility for themselves. Firefighting is dangerous. We all know and accept the risk; that's our job. The charges are unfounded and outrageous. I am not responsible." Pick your excuse. The fact is that Brian's death could have happened in any fire department, including yours. It is only by fate or the grace of God that it did not happen to you or me.

Our dysfunctional fire service seatbelt culture is the root cause of Brian's death. That culture ignores safety standards, does not use readily available equipment, flaunts SOPs, and denies responsibility at the individual, team, and organizational level. Firefighters and officers have told me that using the seatbelts will slow them down, resulting in people dying in fires. The latest comment from a firefighter was, "The only reason we have a seatbelt policy is to cover the department's ass if I get hurt."

Culture drives behavior. Our seatbelt culture let Brian down, and he paid the ultimate price. We are all part of the seatbelt problem and solution. We all must take some responsibility for our brother Brian's death. The question is, what are we going to do about it so it does not happen to another brother or sister?

The National Fallen Firefighters Foundation has just received a one-million-dollar grant to conduct the Firefighter Life Safety Initiatives Project. At its core this project is trying to change our overall safety culture with training programs, lectures, a website, conferences, research, excellence awards, and demonstration projects. One of the U.S. Fire Administration's operational goals is related to reducing by twenty-five percent the loss of life of firefighters. The USFA also uses training programs, publications, research, grants, and lectures. These two national activities are important to changing our overall safety culture. But if we cannot change our seatbelt culture immediately, we have no chance of fixing any firefighter safety issue.

Culture is the collective knowledge, teaching, beliefs, values, feelings, science, technology, art, and behaviors of a group or society. Overall the fire service has a proud culture of professionalism, volunteerism, duty, bravery, camaraderie, and service to humankind. Society holds the fire service in high regard. One national poll reported that society trusted firefighters second only to their own family members. That trust includes returning firefighters to their families after the alarm—unharmed. Firefighters trust each other with their very lives. As professionals, both career and volunteer, we have a lot to be proud of.

Culture is a powerful human motivating factor, and changing a culture is difficult. At its ultimate, a society defines its culture by where it draws the line in the sand. Crossing the line has significant consequences. Sometimes an aspect of our culture must be changed for the good of all. Today's popular phrase is "zero tolerance." For example, our culture of drinking alcohol then driving was changed by Mothers Against Drunk Driving because its members took a stand, drew a line in the sand, and put leadership on the line.

Changing the fire-service safety culture is a big challenge because firefighting is inherently dangerous. But changing our seatbelt culture is doable now. We have the knowledge, we have the equipment, we have the standards, and we have the moral obligation. But are we willing to draw the line in the sand, are we willing to put our collective leadership on the line? Are we willing to impose zero tolerance on seatbelt use?

Fifty-five percent of us wear our seatbelts, but that leaves forty-five percent who do not. It is human nature for some members of an organization not to follow rules voluntarily, so they need to be coerced formally or informally to meet the standards. Let me illustrate.

Formal coercion, for example, is expressed in punishment for breaking the rules. If you are a career firefighter and are late for work, there are consequences for your behavior. The rules and consequences are understood by all. They go something like this. Show up for work on time or this will happen: first offense, a letter of reprimand; second offense, days off without pay; habitual lateness, employee assistance program counseling; and continued lateness, termination. The volunteer fire service has participation standards (drill, meetings, responses, fund-raising, activities, etc.) that members must meet to be a volunteer. If they fail to meet the participation standards, consequences look something like this: extra duty, suspension from riding the apparatus, suspension from the fire station, and termination from the department. Informal coercion, also called peer pressure, is how you treat those who are not conforming. The person is shunned by the group, given a bad reputation, not included in social activities, and counseled by others to do the right thing to be part of the team.

I will bet your career department has over ninety-nine percent of its members report to work on time. Your volunteer department has over ninety-nine percent of its members meet their participation requirements. This is likely the case, because our culture dictates that good firefighters show up on time and meet their participation requirements. Rarely do the formal consequences need to be used, but in certain circumstances, they have been. So we all know where the line in the sand is and that leadership will take a stand. The informal consequences help people conform because they want to be part of a winning team.

What percentage of seatbelt compliance do you have? If the fire service is going to fix the seatbelt problem, we must change our culture by making seatbelt rules and the accompanying consequences more important than attendance. No one dies if we are late for work; no one dies if we miss the meeting;

we do die without our seatbelts on. We must draw the line and take a stand because good firefighters wear their seatbelt.

So, here are the new national formal and informal consequences for not using seatbelts; they go into effective immediately:

- A firefighter not wearing a seatbelt when the apparatus moves will be suspended for one shift without pay (career) or suspended from the fire station for one week (volunteer). The officer in charge will receive the same discipline. The other firefighters on the apparatus will receive the same discipline.

- If a firefighter is injured as a result of not wearing a seatbelt, the firefighter, officer, others on the apparatus, and the fire chief will be suspended for thirty days.

- Firefighters and fire departments that do not comply with their seatbelt rules will not be thought of as good firefighters or good fire departments.

- When the next firefighter is killed because the seatbelt was not used, all of us will cry and feel ashamed; we failed our brother or sister because we did not put our leadership on the line and make them buckle up.

You can change your seatbelt culture or you can think Clark has lost his mind. Either way, I am sitting here crying for Brian Hunton.

Twelve Deaths This Year: It's Time for Seatbelt Hardball
2007

TWELVE FIREFIGHTERS HAVE died in the line of duty since January 2007, in crashes without having their seatbelt on. This must stop. We have no excuse. Almost fifty percent of firefighters nationwide do not use their seatbelt. In some fire departments over ninety percent do not buckle up. Firefighting is dangerous; not wearing your seatbelt is foolish, unprofessional, and in most states illegal. When you don't put your seatbelt on, you dishonor all the firefighters who died to get seatbelts put in apparatus. Shame on us if we do not fix this problem now.

The word *hardball* usually means that something is serious and effort is being exerted to make change by some type of force. The National Highway Traffic Safety Administration's "Click it or Ticket" campaign is just that. The NHTSA make seatbelts a priority for every officer and police agency in the country to give out tickets to everyone who does not have a seatbelt on. Just the threat that the seatbelt law will be enforced with a fine if you are caught not buckled up has pushed seatbelt use to an all-time high of eighty-nine percent nationwide. With the help of a ten-million-dollar national advertising campaign and tickets with a fine, the NHTA has played seatbelt hardball.

The National Fire Service Seatbelt Pledge campaign is not hardball. First, there is no money for ads or staff or websites or material. Second, there is no federal agency in charge of firefighter safety, much less enforcing the seatbelt rules. Third, the fire service, as a culture, does not believe it needs to use seatbelts. Fourth, most fire chiefs and fire officers do not enforce their own seatbelt rules. Finally, too many firefighters believe that real firefighters cannot use seatbelts because another company will steal their fire or a civilian will die because the firefighter cannot get out of the apparatus quickly enough to save them. (These are actual excuses by firefighters.)

So far more than forty thousand firefighters nationwide have signed the seatbelt pledge. Over eighty fire departments have one hundred percent participation in the program and will get a certificate signed by the United States Fire Administration, United States Fire Academy, International Association of Fire Chiefs, National Volunteer Fire Council, National Fire Protection, and National Fallen Firefighters Foundation. The program objective is one million signatures. The impact objective is that no firefighters will die in 2008 because they did not have their seatbelts on.

The Chesapeake, Virginia, Fire Department, under the direction of Deputy Chief E.E. Elliott, is leading the fire service in demonstrating the commitment to firefighter safety when it come to changing our seatbelt culture to ensure that everyone goes home. Some times it takes hardball.

Chief Elliott in his own words

During the very first National Firefighter Safety Stand-Down, our Department [Chesapeake Fire

Department] focused its dialogue specifically toward encouraging and enforcing our existing polices regarding the use of seatbelts. During that year's safety campaign, we initiated an internal Safety Pledge Campaign of our own. Our members were asked to make a pledge that they would never ride in a vehicle without first buckling their seatbelt. Each member making the pledge received a special Safety Award Certificate signifying the commitment to driver safety and the safety of the members of their fire company.

This policy has been aggressively enforced over the last three years. Our safety officer repeatedly forwards LODD notifications, news articles, photographs of vehicle crashes, and eye-opening video clips to all our members to re-enforce and maintain a high level of safety awareness. After three years of aggressively pushing "change," developing and invoking a solid driving policy, as well as encouraging and mandating the use of seatbelts every time a member sits in the seat, one would think this department had attained one hundred percent policy compliance. I did.

Earlier this year, however, I was returned to the real world. Within about the same timeframe, two red flags were raised. First, our department's Safety Committee was receiving complaints from some company officers that while filling in for other officers, they felt like they were the "bad guy" for enforcing the seatbelt policy when, apparently, the members of those particular crews had not been adhering to it. The second red flag, and perhaps the most eye-opening revelation, came about during our biannual promotional process. While sitting in the fire chief's interview portion of the process, there were candidates who took the absolutely correct position on "leadership by example," and at the same time, openly admitted that they either 1.) did not always wear seatbelts themselves or 2.) did not always ensure that their crew was belted in. I can commend the candidates' honesty; however, their responses were the warning signal that we were not where we thought we were.

One of the sixteen Firefighter Life Safety Initiatives now in place, actually the very first one regarding cultural change, places responsibility on leadership. In response to this initiative, we now have a catchphrase aimed at firefighters: "The courage to be safe!" The question should be turned around and pointed to chief officers and fire-service leadership. "Do we have the courage to be safe?" Inevitability, our actions speak louder than our words.

My personal commitment to the 2007 Firefighter Safety Stand Down was to ensure, to the best of my ability, that "Everyone in the Chesapeake Fire Department Goes Home!" To that end, I raised the stakes on non-adherence to our standing driver-safety policy mandating the use of seatbelts.

Effective July 1, 2007, the violation of our policy is now in the most serious category of policy violations: "violation of safety rules where there is a threat of bodily harm." With this revision, non-compliance can now result in the termination of employment for the individual, the driver, and the officer. (This is seatbelt hardball!)

As difficult as this action feels for me personally, it is so easy when compared to attending a LODD funeral. Even as we have worked in the past to change fire-service culture toward discrimination and harassment, some facts remain the same. There are members of the fire service who will not "get it" until somebody loses their job; and those firefighters that still don't "get it" should not be associated with the fire-service organization.

The rest of the story

On July 3, 2007, two days after the stakes were raised on the mandatory use of seatbelts in our department, we experienced an unfortunate, but perhaps timely, incident.

While traveling on a back, rural road, the freshly paved asphalt road surface broke away under the weight of the fire apparatus, plunging the pumper headlong into the ditch. (There was absolutely no road shoulder.) The engine proceeded to plow the ditch for approximately sixty feet, taking out a telephone pole in the process. Eventually, it came to rest almost on its side with a grass fire now burning close by due to downed power lines. All four members were securely buckled in at the time of the incident and were able to extricate themselves from the engine, walking away with only minor cuts and bruises.

During my personal interview with the company captain, he stated, given the impact to which he was subjected, he shuddered to think what the end result would have been if any one of the four had not been secured in the cab with seatbelts.

Chief officers, there are two questions you need to answer regarding the use of seatbelts in your departments. First, do you know the differences that exist between your policies and your practices? Moreover, do you have the courage to do what it takes to bring them into alignment?

The Chesapeake Fire Department experience is an excellent example of "Talk softly but carry a big stick." When it comes to firefighter seatbelt safety we must hold ourselves and each other accountable. Take the pledge . . . buckle up . . . enforce the policy . . . make the discipline hardball. The alternative is deadly. Firefighters and fire chiefs you choose!

"Click it or LODD."

Our Seatbelt Tale
2007

> It was the best of times, it was the worst of times, it was the age of wisdom, it was the age of foolishness, it was the epoch of belief, it was the epoch of incredulity, it was the season of Light, it was the season of Darkness, it was the spring of hope, it was the winter of despair, we had everything before us, we had nothing before us, we were all going direct to Heaven, we were all going direct the other way—in short, the period was so far like the present period, that some of its noisiest authorities insisted on its being received, for good or for evil, in the superlative degree of comparison only.
>
> –A Tale of Two Cities, 1859

CHARLES DICKENS'S OPENING sentence is a statement about all time, all people, all conditions, all societies, and all issues past, present, and future. The statement captures our human condition: the alpha and omega, the yin and yang. Our free will is a gift to use between the beginning and the end, the positive and negative; how we use it is up to each of us individually and all of us collectively. Dickens's sentence represents the results or outcomes of our actions, which lead to the tales we tell.

So, what is our seatbelt tale for 2007?

Best: Fifty thousand signatures on the national seatbelt pledge and ninety-one fire departments achieve one hundred percent participation.

Worst: Twelve line-of-duty deaths due to no seatbelt. We have averaged ten per year for the past thirty years.

Wisdom: Five fire-service organizations have put their logo on the one hundred percent seatbelt pledge certificate.

Foolishness: One major fire service organization's logo is not on the certificate.

Belief: The fire service has the knowledge, skill, and ability to fix our seatbelt problem.

Incredulity: Not wearing seatbelts is still our number-one safety violation.

Light: Chiefs, officers, and drivers are responsible for everyone to buckle up.

Darkness: Chiefs know the seatbelt SOP is not followed and do nothing about it

Hope: No line of duty death in 2008 because of no seatbelt and one million pledge signatures.

Despair: What family will lose their loved one needlessly?

Everything: The National Fallen Firefighters Foundation seatbelt seal of excellence is waiting for the fire service to earn it.

Nothing: Can bring back a fallen firefighter.

Heaven: Smiles on seatbelt use.

The Other Way: We all know what happen to those who do not follow the commandments.

In short, our seatbelt tale has not changed for the better, and this is sad. Next year, are we condemned to repeat the past? Will ten families get the terrible visit by the fire chief? Who will the chief be? Will our "noisiest authorities" insist on seatbelt use by firefighters, or will the future be like the past? It can't happen to me, it can't happen to anyone I know, it can't happen to my apparatus, it can't happen to my crew, it can't happen to my fire department. Seatbelts are not a priority for us.

Our 2008 seatbelt tale will be written by you: Take the pledge, buckle up, tell your buddies to buckle up, do not move the apparatus until all are buckled up, and enforce the seatbelt rules. It will make a great story to tell your family: "I wear my seatbelt on the fire apparatus because I love you. All firefighters do, so Everyone Goes Home."

The Princess, the Governor, and the Firefighter
2007

WHAT DID THE princess, the governor, and the firefighter have in common?

They did not wear their seatbelts. The princess was known worldwide, but she did not have her seatbelt on when the car crashed; she died. The governor did not have his seatbelt on when the car crashed; he was severely injured. Since the first National Fire Service Safety Stand Down Day in June 2005, too many firefighters have been killed and injured in department apparatus because they did not have their seatbelts on.

The number-one safety violation committee by firefighters is still not using their seatbelts. This means *we are not* "Ready to Respond." If we know this, why can't we fix it?

A colleague of mine who investigates line-of-duty deaths went to visit her daughter and grandson. My friend had just completed a seatbelt-related LODD investigation. The crash was still on her mind when she met the six-year-old grandson and he asked her why she was sad. She explained that a firefighter had died because he fell out of a firetruck. The boy asked her if he had his seatbelt on, and she replied that he did not.

"Why not, Grandma?" he asked. "Well, it's complicated," she said (referring to the explanation not the act). He looked at her with a puzzled look and said, "No, it's not, Grandma. It's real easy; you just click it together!" The six-year-old, the princess, the governor, and the firefighter all know. "It's real easy [to use seatbelts]; you just click them together." But, do they buckle up every time? The only one I am one hundred percent confident in is the six-year-old.

Does the fire service want children to see us riding in firetrucks without our seatbelts on? Does your governor know that firefighters in the state do not use seatbelts all the time? Does your mother, spouse, children, and grandchildren know that you do not use your seatbelt when on the firetruck? Do you know that firefighters died to get seatbelts put on your fire apparatus? We can fix our number-one safety violation immediately.

Has your fire chief asked you to take the National Fire Service Seatbelt Pledge? Have the state fire chiefs and firefighters associations asked its members to take the pledge? Can one million firefighters promise to wear their seatbelts? Can we follow our own seatbelt rules? I hope your answer is *yes* because it is easy to take the pledge and to buckle up.

If your answer is *no*, you can start planning for more funerals and visits to hospitals because you are not ready to respond.

Thank you to all the firefighters who have taken the pledge. Congratulations to all the fire departments that have one hundred percent participation in the seatbelt pledge campaign. Your efforts are making a difference to help ensure that Everyone Goes Home. Whether you're a member of the royalty, a high elected official, a firefighter, or a six-year-old, we all can make a difference one click at a time.

LODDs: We Will Forget You!
2008

DO YOU REMEMBER the twelve seatbelt deaths from 2007? We pride ourselves on the slogan "We Will Never Forget," but our behavior speaks louder than our words.

Twelve firefighters died because we did not make them put their seatbelts on. History repeats itself, according to Chief Billy Goldfeder of the The Secret List:

> **Detroit Firefighters Ejected—Again ... Close Call**
> As you should remember, on 2-7-07, Detroit Fire Engineer Joseph Torkos of Engine Company 17 was tragically killed in the Line of Duty after he was ejected from the apparatus following it being struck by a speeding SUV.
> And now this morning, 1 Detroit Firefighter needed stitches in his head and another has a broken arm after their rig flipped while on their way to a fire this morning. 2 other firefighters were not seriously hurt when the truck skidded and rolled onto its roof at 0715 hours this morning.

Witnesses said they saw all 4 Firefighters thrown from the apparatus, although the rigs are equipped with seatbelts. One Firefighter was nearly crushed, a witness said.

Every safety rule and every piece of safety equipment was paid for in blood by firefighters who came before us. When we do not use the equipment and follow the rules, we are disrespecting our dead and injured brothers and sisters.

But we can change at the individual, company, and department level. There are 58,700 fire-service leaders who are making a big difference by taking the National Fire Service Seatbelt Pledge, buckling up, and not moving the apparatus until all on board click it.

Lieutenant Lauren Brown of the Dallas Fire Department is one of them. This is her story in her own words:

> May 14, 2008
> Dear Dr. Burton Clark:
>
> Here is my "seatbelt story." In short, it is a brief account of how I came to realize that I was not doing my job as an officer or as a firefighter.
>
> I'm not sure which of the following events actually happened first, but they both were within the first two days of my Interpersonal Dynamics class. One morning, you stopped by to introduce yourself and to raise awareness about the Brian Hunton National Fire Service Seatbelt Pledge. At about the same time, Fire Chief Tom Taylor from the Moses Lake Fire Department in Washington asked those of us in class to sign the Pledge.
>
> There are a lot of "firsts" in this story... This was my first time to attend the NFA, the first time I had heard about Brian Hunton (even though he was a fellow Texas firefighter) and the first time I

I Can't Save You But I'll Die Trying: American Fire Culture

had heard about the Pledge. And finally, I signed the Pledge without giving it the first thought as to how it would impact my station life and my duties as a fire officer.

My first two weeks at the Academy were filled with sleepless nights as I lay awake wondering how I was going to make this change for myself, my crew, and my department. I took the opportunity to reflect on the many inconsistencies in my life that were related to what should have been such a simple, and automatic, action.

I had to admit that I had overlooked our department's policy, and state law, for my entire career as a firefighter and an officer. I wore my belt in my personal life, and not in my professional one. In fact, when I was growing up, my parents would fine me a dollar for every time I did not buckle up. At work, I wore it religiously when I was on the ambulance, but once I slid my gear across the apparatus room floor to ride the engine or the truck, I was never buckled.

After signing the Pledge, I called home one night and told my husband to "get ready" to wear his seatbelt once I came home; this was his two-week notice. He quickly came up with a variety of excuses that all boiled down to either 1) it's too hard to change old habits or 2.) we don't buckle up en route to a structure fire because we are trying to get dressed and provide the fastest response. This is my husband, also a Dallas firefighter, who had just left the station about two years ago when the engine he was riding rolled onto its side. (I remember the undocumented reports from the scene talking about gear moving around the cab like a hamster in its exercise wheel.) The thought that I could not even convince my husband to buckle up was, at once, discouraging and a challenge that I could not let go.

Thank goodness firefighters are competitive people. I am not sure that Chief Taylor knew it, but I felt accountable to him; he was one of the few in my class that I told of my goal to make Dallas Fire-Rescue one hundred percent compliant. And, now that my husband had told me "no," I was certain that I would make the Pledge a success.

I knew that once I returned to my station, my work was cut out for me. First, I had to make sure that I wore my seatbelt, every time. Then, I talked with my station captain and we talked to our crew. We all signed the pledge for our New Year's resolution at the table after breakfast one morning and I passed out "Everyone Goes Home" helmet stickers. Next, I told my battalion chief of my plans to do the Pledge department-wide and he put me in contact with our fire chief.

I solicited support from members at all levels, including some operations chiefs and our Safety Advisory Committee. I coordinated two roadway safety training lessons for our members and administrators and pledge copies were distributed according to assignment and shift. In less than three months, we have obtained 1,748 signatures from our uniformed and civilian employees who are eligible to drive or ride in fire department vehicles. Last week, Fire Chief Eddie Burns certified that we are one hundred percent compliant.

The change for us has been gradual, but make no mistake, it is deliberate and a priority. I still check myself and my crew on almost every emergency run and even on our return trip home.

However, they usually beat me to it and are the first to scream "Seatbelts!" in their craziest voices as we run out to our rigs.

I truly believe that this change will save lives and I am ashamed that you had to bring it to my attention. The power of peer pressure, especially firefighter initiated peer pressure, is overwhelming. I hope that we can use a little friendly competition to our advantage and generate change among everyone, especially the large metro-sized fire departments, so please pass the word that Dallas did it!

Sincerely,
Lauren A. Brown

I hope there is an officer like Lieutenant Brown in every fire department. Detroit owes it to Joe Torkos. The rest of us owe it to the three-hundred-plus firefighters who have been killed in crashes over the past thirty years because no one made them put their seatbelts on.

Chief: Does your fire department have a one hundred percent seatbelt pledge certificate signed by U.S. Fire Administration, International Association of Fire Chiefs, National Volunteer Fire Council, the National Fire Protection Association, and the National Fallen Firefighters Foundations? If not, why not?

The worst day of your life will be when you go tell a mother, father, wife, husband, or child that their firefighter will not be coming home because you did not make them put their seatbelt on. You may also want to tell them, "We will forget them soon."

Two Seatbelt LODDs Remembered
2009

IT IS FIREFIGHTER Safety Week. The slogan this year is "Protect Yourself: Your Safety, Health and Survival Are Your Responsibility." In 2008 we lost ten firefighters to no seatbelt; in 2007 we lost twelve. How many will we lose in 2009?

Over three hundred firefighters have been killed in the line of duty over the past thirty years because they weren't wearing a seatbelt. We are still averaging ten seatbelt LODDs per year. Of all the ways a firefighter can be injured or killed, not wearing a seatbelt is the most tragic because it is one-hundred-percent preventable.

Over 107,000 firefighters have taken the National Seatbelt Pledge, and over five hundred fire departments have achieved one hundred percent participation. The North Carolina fire service is leading the nation with 159 organizations that have one hundred percent.

The objectives of the Seatbelt Pledge campaign are one million signatures and thirty thousand fire departments with one-hundred-percent participation. The goal is one whole year with no LODDs due to no seatbelt being used.

When you conduct your safety week activities have all you members read the following two stories. I asked my friends Stan and Richard to write these stories because they may prevent another tragedy.

Collateral damage
By Stan Lake, Deputy Chief (ret.), CAL FIRE/
Riverside County Fire Department

Where were you when . . . ? The major events in history that affect the nation or the world are all easily remembered. We can easily recall where we were and what we were doing. In the fire service, a line-of-duty death of one of our firefighters can generate those same memories.

If we see it on the evening news, read about in a newspaper, or get more details from Billy G's "The Secret List," we can feel a sense of loss for those other departments. We are sorry, but are also glad it didn't happen here to us (or to me). And then, it did.

My wife and I went out to lunch on that day in August 2005. While enjoying a quiet afternoon at a favorite restaurant, I noticed the weather outside had changed from distant clouds to dark skies and intense winds. Clouds of dust and loose debris were swirling down the street past the restaurant. Other customers were remarking about this unseasonable sudden storm. I silently hoped the wind would be accompanied by rain—strong winds in August in southern California are not a good thing.

As a deputy chief with CAL FIRE/Riverside County Fire Department, my cell phone didn't ring often on the weekend. However with this wind event, I wasn't surprised when the call came. It was one of my battalion chiefs. His report was not what I expected. "Chief," he said, "Engine 58 has been in an accident. It sounds bad. There are injuries." I requested more details. The battalion chief said he was en route and would be on scene in a couple of minutes. His next report carried my worst fears. Two firefighters had been ejected from the engine. One of them had ended up under the engine. He was being transported to the nearest hospital with major injuries. I questioned the battalion chief about the firefighter—who was it and how severe were the injuries. "It's Chris Kanton. We just need the doctor to tell what we already know. He didn't survive."

My emotions were simply numb. I couldn't say anything for a moment. Then the years of experience kicked in, duty called, and I went to work. I coordinated with the chief of the department and other members of the executive team for the things that must be done. While others would make family notifications, work with law enforcement agencies, and initiate our own investigation, I would go to the hospital that was receiving the other injured firefighters.

When I arrived at the hospital emergency room, both firefighters were being examined and treated. I soon learned the firefighter who had not been ejected had only minor contusions and abrasions. The other firefighter—the operator of the engine—although conscious and coherent, had suffered unknown, possibly severe head injuries. The doctors wanted to transport him to another facility better-equipped to treat his injuries. I asked if I could see him and was given the okay.

When he saw me, he started to cry. "They won't tell me about Chris." I looked at him and knew I had to be honest. "Chris didn't make it." I stayed with him for a while and we talked. Talked, not about the accident, that would happen soon, but about his taking care of himself and doing what was necessary to get better. I saw to the other firefighter's needs and ensured he had family there to be with him until the doctors released him.

I went out of the ER to see the other firefighters who I knew had been gathering. There were several of them waiting for news. By this time, they already knew about Chris. I looked into their faces and saw a variety of emotions. Some showed fear, some anger, and some grief, others simply stoicism.

There were hugs all around. When I started in the fire service, hugging was not common. I'm glad that has changed. I felt their pain as I tried to stay calm for them and me. I stayed with the injured firefighter until he was transported to the second hospital. He went by helicopter, I followed by vehicle. It was determined he had suffered a concussion and severe lacerations to his head and ear. His injuries were not life-threatening, but he would be hospitalized for a few days.

Finally, I was able to start to deal with what caused the accident and how the injuries occurred. The accident was caused by a sudden, severe rain storm that made the highway extremely dangerous to negotiate. The engine operator lost control of the vehicle, which collided with the center divider and careened off the roadway and down an embankment. It hit two trees before coming to rest on the roadway below the freeway. As the engine spun out of control, Chris (riding in the back-facing open cab) was ejected. He was subsequently run over by the engine, causing fatal injuries. When the engine hit one of the trees, the operator was ejected through the windshield. The third firefighter remained in his seat. How had the one firefighter not received serious injuries while the others had? I, of course, already knew the answer. Seatbelts don't fail—unless they're not being used.

Chris's funeral was everything you have heard about those events. The bagpipes, the procession, the ringing of the last alarm, the eulogies—these all were happening in a blur to me. I was still numb. I went home. I was numb. I could think of little else but Chris. He was one of my "boys." Finally, I just cried. And cried...

I realized I needed to talk to someone about my feelings—feelings that *I* had somehow been responsible. My wife knew I was having problems and was very supportive. However, professional counseling was what was needed. I have been very fortunate to have an outstanding counselor. It has taken quite a while for me to really deal with those events.

Burton A. Clark

The engine operator also had an ordeal to cope with after the accident. As is normal, he was placed on administrative leave pending the outcome of the investigation. He would not return to work for almost two years. After the accident investigation, he was charged with vehicular manslaughter. The charge was eventually dropped when weight-distribution factors in the engine's design were shown to be the cause of the accident. Although exonerated, he will always carry the memories of that terrible day.

I have always been very concerned about the safety of firefighters. Obviously, it is an inherently dangerous job. The fire service spends large sums to ensure the safety of its members from all manner of hazards found on the job. It has always seemed natural to me to buckle up when in a vehicle. I'm old enough to remember when cars didn't come equipped with seatbelts. I helped my dad install them in our car. I assumed everybody else used them just as I did. As is often the case, making assumptions can be bad for everyone concerned.

I've talked to many firefighters about seatbelts and their use. I found most firefighters do use their seatbelts. However, there are many (too many) who don't. There are several reasons cited when this practice is rationalized. We've heard them all over and over again. In the minds of these individuals, the possible consequences are, at best, minimized. "Let's just jump on the rig and get to the call." As leaders, it is our responsibility to demand the safety of our people. We talk about "everyone goes home." You can't return from a call if you don't get there in the first place.

I am glad to see the memory of Chris Kanton goes on. He can be found on the internet. His name is on a plaque at the National Fallen Firefighters Monument in Emmitsburg. A portion of an interstate highway and a fire station have been dedicated to his memory. His legacy should be an inspiration to all. His life, as short as it was, was a full one complete with all the dreams a firefighter with tremendous potential can have. His life is still inspirational to those who knew and loved him, as well as those that simply hear about him. Hopefully, his memory will also prevent the loss of another firefighter who doesn't think the use of a seatbelt is that important. Hopefully, his memory will prevent the grief that must be endured by another family, by other friends, other co-workers, and by other chief officers charged with firefighter safety. Collateral damage is what can happen to those close to a serious incident. I never thought it could involve me. Don't let it happen to you.

John Francis Keane, another senseless loss to the Fire Service
By Richard Hart, Battalion Chief, Waterbury, Connecticut

Incident #07-2224 seemed innocuous enough, another call for a kitchen fire, of which there were 175 for 2006. The standard response, three engines, a truck, the rescue, and a chief officer assigned to the box proceeded as usual. At 1033 hours on May 19, 2007, the world as it was known to the Waterbury Fire Department changed forever.

I was at my son's hockey game when, as oftentimes it does, my cell phone rang and the familiar voice on the other end asked if I had heard of an accident in Waterbury involving apparatus that morning. I said, "No, but I will find out and call you back." I immediately called our dispatchers and received the news that Engine 8 and Truck 1 were involved in an accident, and John and Joe Fischetti were "critical." I told the coach and my son I had to leave, and I made the fifteen-minute drive to Waterbury Hospital ER where I was met by other off-duty firefighters. From the looks on their faces and the tears, I knew the next few hours were going to be the most challenging and gut-wrenching of my career. It was worse. I received an update from the charge nurse whom I had known from my years as a paramedic, and she informed me John was to be transported to Yale-New Haven Hospital's

Neurological Intensive Care Unit. With that I knew the outlook was grim at best.

We, as firefighters, oftentimes view promotional exams or large, complex incidents as "the most challenging moment of our careers," yet facing the wife and children of a brother who lies dying in a hospital bed takes the challenges we face to a new level.

Monica sat quietly in the family room with a priest and two nurses and when I entered, I was amazed at the peace and serenity that surrounded her. I relayed the information about John's being transferred to Yale and that I would escort her down. The chief of department came in to offer any assistance and permanently assigned me to Monica throughout this ordeal.

That Saturday afternoon in May, John was transported to Yale's NICU, one of the world's most renowned hospitals. Monica, John's sister, Maura, and I followed the ambulance for the forty-minute drive.

The muted silence of Yale's NICU was a sharp contrast to the pandemonium of the emergency room at Waterbury Hospital. The head of trauma handled John's case and took time to explain in detail what John faced and the extent of his injuries. In the back of my mind, I knew John died that Saturday morning. For three days, I and Lieutenant Bob Wall, John's and my good friend, stayed at the hospital to watch over John and provide support for Monica and the rest of John's family on a rotating basis. Neither of us were away from the hospital for more than five or six hours, just enough time to reacquaint ourselves with our families and grab a few moments of sanity. New Haven firefighters came and went, bringing bagels and sandwiches for us; the hospital provided our department with a large lounge to "live" in for the duration of John's stay. Unfortunately, the hospital was well-versed in providing support for the families of public safety personnel; this was another instance of its compassion that went beyond the medical aspect of care.

The days dragged, and on May 22, I brought Monica to her home and I went to my home to visit with my wife and three children. At 2300 hours, the phone rang and the person on the other end of the line simply stated, "John is trying to die." After a flurry of questions and some confusion, I left for New Haven.

As I was driving, I phoned Bob and explained the situation and that he needed to get to Monica's house. I arranged for the closest engine company to be at the house to watch the kids until Bob and Monica's mother got there. Bob got to the house and drove Monica to the hospital.

John succumbed to his injuries on May 22 at 0205 hours surrounded by his wife, brothers, and sisters. The irony was that I knew John was gone on May 19, the day of the accident, although the finality of the pronouncement by the doctor was gut-wrenching. I had accompanied John down to the nuclear CT scan required to confirm his death. Monica couldn't bear to be there, so I felt it my duty not to leave his side. At the conclusion of the test, the nurses could not tell me the outcome, but I asked anyway. From her expression I knew the answer, then she said how sorry she was. I could not cry; I still had to fulfill my duty.

When we returned to the NICU, Bob looked at me and knew as well. I went out of the unit to announce John's passing to a group of approximately twenty firefighters.

The wake and funeral were a blur. Bob and I never left each other or the family, except to sleep at home. Monica asked me to write the eulogy for John, which I accepted. I struggled with the content, only because John was truly a shining star, destined for greatness, and was unsure as to the direction to go in. I wrote to his kids, nothing more, nothing less. I told them what a great person their dad was and to never forget the good times they had together. John was one of the most honorable and moral persons I have known. I could not cry—too many people needed information and talks on seatbelt safety and John as a person. My wife was my rock throughout this whole ordeal. She comforted me and gave me words of wisdom and compassion, and for that I will always be in her debt and love.

John was my best friend on the department; he was initially assigned to the shift opposite me, but we worked together constantly. We sat many nights together discussing various issues and dilemmas facing our union, department, and city. We had many arguments, but we always agreed on one thing: We loved our job. John was destined to be a firefighter, following in his the footsteps of his grandfather, who retired as a battalion chief. John was a good firefighter.

These events, or very similar ones, play out across the country approximately thirty times per year. It is time we stop killing ourselves in the senseless, selfish manner of failure to wear a seatbelt. The attitude of our profession must change; we are not bulletproof or indestructible. Take the National Seatbelt Pledge and wear your belt. Live the pledge; do not put your loved ones through the agony of your senseless death; do not leave your children without their father, your wife a widow.

It has been two years, and much has changed within our department, mostly for the good. There is a John Keane Memorial Golf Tournament every September to raise scholarship money in John's name for elementary school children. The National Seatbelt Pledge was taken by all members, and safety and seatbelts have taken on a new meaning. Policies were updated and finally enforced. The NIOSH report and police report confirm what we already knew—John was not wearing his seatbelt.

I miss John, and I still have not cried.

* * *

Thank you, Richard and Stan, for sharing your loss with us. Your experience is one none of us wants to have. Who is responsible for seatbelt us in the fire service? All of us. Take the National Fire Service Seatbelt Pledge get your department a one-hundred-percent certificate. It is in our power to make sure we all buckle up so everyone goes home.

We did not reach out goal in 2009. Will we in 2010?

The Role of Leadership and Seatbelts
2009

THE STORY OF four leaders from the State of North Carolina; the Waterbury, Connecticut, Fire Department; the Chief of the New York City Fire Department; and a West Point cadet will be presented in this article. These four examples may help guide the rest of us because that is what leaders do—they show us the way.

The State of North Carolina has two fire departments that have achieved one hundred percent seatbelt pledge participation. This accomplishment is because the North Carolina fire-service leadership made a commitment to change its seatbelt culture. The state firefighters and fire chiefs associations made the seatbelt pledge campaign one of their top priorities. Across the state, from large career departments to small volunteer departments, from firefighters to fire chiefs, everyone gets the same message: Take the pledge and buckle up. North Carolina is ranked number one in fire departments with one hundred percent seatbelt pledge participation; it is showing the way for the other forty-nine states and the District of Columbia.

After the line-of-duty-death of Captain John Keane in an apparatus crash in 2007, the Waterbury Fire Department wanted to make a change. Wearing seatbelts became a number-one priority for all firefighters. As a department they did not want Captain Keane to have died in vain. The entire fire department took the National Fire Service Seatbelt Pledge to honor John, to commit to each other that everyone will wear their seatbelt every time, and to send a message to the nation's fire service about the importance of seatbelts so other fire departments can avoid such a tragic loss.

When all fire chiefs make seatbelt use a priority and hold everyone accountable, our seatbelt line-of-duty deaths will stop. FDNY Chief of Department Salvatore Cassano has put his leadership on the line when it comes to seatbelts. The FDNY safety culture video was premiered at Firehouse Expo 2009, and part of that cultural change is getting firefighters to buckle up. During a Radio@Firehouse podcast Chief Cassano signed the National Fire Service Seatbelt Pledge and plans to send it throughout the department. Chief Cassano is showing all metro fire chiefs the way to get firefighters to buckle up.

The final story will be told by a West Point cadet. The U.S. Military Academy is all about making leaders for the Army and our nation. The fire service's dysfunctional seatbelt culture is very powerful; it is so powerful that it let Cadet Lewis Han disobey a direct order. West Point cadets do not lie, cheat, steal, or tolerate those who do, so the cadet was required to tell this story. Cadet Han learned a valuable lesson from his experience riding on a firetruck. What he learned is a reflection on us. It is up to every fire service company officer to learn from this story.

Seatbelts and leadership
By West Point Cadet Lewis Han (D4 2010)

"Research has shown that lap/shoulder belts, when used properly, reduce the risk of fatal injury to the front-seat passenger car occupants by forty-five percent and the risk of moderate-to-critical injury by fifty percent. For light-truck occupants, safety belts reduce the risk of fatal injury by sixty percent and moderate-to-critical injury by sixty-five percent."

I recently had the opportunity to ride along with a fire department engine company. It was part of an internship in which I was taking a look at the need for management skills and core competencies (written and oral communications, interpersonal skills, group and team skills, and organizational skills) in fire officers as well as any other management position. I was able to ride along with a particular fire engine company for a full twenty-four-hour shift. Some of the things I learned, especially the unexpected lessons, seem to hold the greatest significance.

Being a West Point cadet with a focus in management, I have always been interested and intrigued by the topic of behavioral sciences. I witnessed the camaraderie that forms and builds between the firefighters. Not only were all the firefighters of this engine company absolutely competent and experts in their field, they also worked as a cohesive unit. My respect for the fire department and its firefighters has grown immensely through my experience as a ride-along. The lieutenant especially impressed me because of the leadership qualities he possessed. He had developed a keen sense of interpersonal and team skills. From what I observed, the firefighters of that particular engine company really did respect him, as did I.

What does any of this have to do with seatbelts? Well, while the experience at the firehouse will leave me with great memories for the rest of my life, there is also something that should be addressed. I have learned that firefighters in general do not wear seatbelts. Many states actually do not require firefighters to wear seatbelts while responding to an emergency. The reasoning behind this argument is the fact that firefighters must be quick and mobile while responding. Therefore, firefighters have built up a culture in which they believe seatbelts "hinder" them from completing their duties in a timely fashion.

   I was instructed that I was to wear my seatbelt as a ride-along even if the firefighters themselves didn't. It would have been tragic if a visitor had gotten injured or killed while riding with the fire engine company. Now, when I think back to the calls, I remember there were times when I did not buckle my seatbelt. This action was usually accompanied by frantic movement inside the truck with firefighters jumping into their fire suits and gearing up for a fire call. Because there is limited space and a cluster of equipment in the backseat, I did my best to help move things out of the way and "think skinny." The fact that I did not buckle my seatbelt every single time I got on the engine is absolutely one hundred percent my personal failure, and I take full responsibility for my actions. I was instructed to do something, and I failed to keep up my end of the bargain.

   However, while pondering the event, I realized there were heavier consequences than just a personal failure. Unfortunately, however much I want for the buck to stop at myself, it doesn't. Any student of leadership will tell you that a leader is responsible for all of the organization's or unit's successes and failures. Consequently, for the time that I was under the lieutenant's responsibility, if a ride-along had been injured, the responsibility would have fallen completely on him. Whether anyone thinks this is fair does not matter. The responsibility to lead and be accountable for me as well as the other firefighters would have been on his shoulders. It is simply something that comes with being in a position of leadership.

   Through this experience I have learned a couple valuable lessons. The first is that as a follower, you must realize that you will not always be able to "take responsibility" for your own actions. Whenever you fail to do something, you let down those around you, and you put your leader in a bad position because he is responsible for you. It does not matter if you kick and scream saying that the leader should not be blamed for your actions; it will happen regardless. The second lesson is that as a leader, you truly do have the responsibility for everything that succeeds and fails in an organization. Because of your position, you have the ability to influence others. The men and women assigned to you should be accountable to you, but you are also accountable to them. In this case, it is the issue of putting on a seatbelt. But the fact still remains for all things, if the leader performs an action and instructs the others to do so it will happen.

Once again, I do take all personal responsibility for the failure to wear the seatbelt, but I know that it is not that simple. And that is what I will take away from this experience forever; the second-order effects of actions and the implications that they hold for those around you. I have learned key lessons and I hope to keep them in mind whenever I hold a position of leadership. Fortunately, this article does not end in a negative fashion. Because leaders are in a position of influence, there is hope. There is hope that the leaders of the fire service will take care of the most valuable asset they possess. It is not the expensive machinery or costly operations. But instead, the most valuable asset is the men and women who show up every day and give their service to the general public. If each individual fire officer ensures that their immediate subordinates are buckled up, the entire organization will be buckled up. I just hope that the firefighters, who work so hard to keep us safe, will keep themselves safe as well.

* * *

Cadet Han is to be commended for taking the time to share his experience with us. I hope the fire service hears his leadership message loud and clear.

If FDNY Fire Chief Cassano can take the pledge and work to get the largest fire department in the country to buckle up, the rest of use have no excuse not to do the same.

For the Waterbury Fire Department to suffer the loss

of Captain Keane and turn his memory into a one hundred percent seatbelt pledge certificate brings new meaning to the phrase, "We will never forget to buckle up."

We all need to thank the North Carolina fire service for showing us what can be accomplished when we want to change our seatbelt culture. Let's see if North Carolina remains number one.

Leadership and seatbelt stories are all around us. You have your own personal seatbelt story. Is your story a seatbelt leadership example from which we can all learn? When it comes to seatbelt use in the fire service, do you show the way?

What Do I Stand For?
2010

THIS ARTICLE IS an apology to a pair of state firefighters and fire chiefs associations; it will describe a significant learning experience I had. I hope it helps the fire service see the light and be the light when it comes to firefighter seatbelt safety. I also hope it gives me the courage to turn vision into action.

I was invited to be the keynote speaker and a workshop presenter at joint annual conference of state firefighters and fire chiefs associations. Naturally my presentations contained firefighter seatbelt use references. After both presentations, I received very positive feedback from several people on the content of my message and my skill at public speaking. I was feeling very satisfied with myself and the seatbelt accomplishments of the state fire service.

This state fire service has moved up the national list of the number of fire departments that have achieved 100% seatbelt pledge participation. In addition, it passed a resolution to petition the state legislature to pass a law that will require seatbelt use by firefighters. Good stuff. Then, I went to the Friday night firefighters parade.

I Can't Save You But I'll Die Trying: American Fire Culture

Dozens of apparatus, hundreds of firefighters, less than thirty percent seatbelt use. One FD pickup truck had eight firefighters standing and sitting in the truck bed; two were sitting on the down tailgate. Firefighters in apparatus crew cabs and front seats: no seatbelts. Firefighters in apparatus crew cabs holding children: no seatbelts. Firefighters standing on the back step: no seatbelts. Firefighters sitting in the hose beds: no seatbelts. Firefighters sitting in hose beds holding children: no seatbelt. If we can't get firefighters to use seatbelts at a parade, how are we ever going to get them to buckle up going to a fire?

I got angry and disappointed; I put my head down, closed my eyes, and shook my head. Then I started to bluster to the fire chiefs I was with that the next day I was going to give hell to the associations for letting this happen. Didn't they know how terrible it would be if a firefighter fell off an apparatus and was killed? Didn't they know what a tragedy it would have been if a child had fallen off the apparatus and was killed?

My anger made it difficult to sleep. In the morning I realized I was not angry at the state firefighters and fire chiefs associations. I was angry at myself, because I saw a wrong and did not stop it. I closed my eyes and let the wrong pass by. I am ashamed of myself for not having the courage to walk out into the middle of the street, put my arms up, and stop the parade.

I did not get a chance to address the associations on Saturday morning. So this will have to serve as my official apology to the state firefighters and fire chiefs associations for getting angry at you. Please forgive me for letting your firefighters and your children be at risk of death and injury due to no seatbelt.

The state's fire service seatbelt efforts are to be applauded. The firefighters and fire chiefs associations are providing the leadership to get firefighters to buckle up all the time, even at parades.

The quote I used in my speech from Joel Barker's Power of Vision video, "Vision without action is merely a dream. Action without vision just takes up time. Vision with action can change the world" is still true and a principle I try to live by. I failed to take action at the parade. I will not make that mistake again. So, if you invite me to a firefighter parade make sure everyone is seated and belted or I will be standing in the middle of the street.

China's "Tank Man" of Tiananmen Square disappeared; some believe he was executed for taking a stand. In the United States of America we have the right and duty to take a stand. Many countries around the world do not have that same freedom. Many of our brother and sister firefighters around the world cannot take a stand on what they believe.

What does the America Fire Service stand for? Standing for firefighter seatbelt use will not get any of us executed, and there cannot be two sides to the value and behavior of firefighters' wearing their seatbelt.

August 2010 marked my fortieth year in the fire service. We have had seatbelts in our apparatus for over thirty years. We have lost over 343 firefighters in the line of duty because we did not make them buckle up.

If we all stand up for seatbelts and take actions to ensure seatbelt use, the vision of no firefighter LODDs due to no seatbelt will become a reality.

It takes courage to ask these questions. What do I stand for? What actions do I take to get firefighters to buckle up? See you at the firefighters' parade.

The Great Firefighter Seatbelt Lie
2013

THE 2013 INTERNATIONAL Fire/EMS Safety and Health Week focused on behavioral health. I thought I understood the construct of behavioral health; I looked it up on Wikipedia anyway.

"In psychology, behavioral health is a general concept that refers to the reciprocal relationship between human behavior, individually or socially, and the well-being of the body, mind, and spirit whether the latter are considered individually or as an integrated whole."

Since my prostate surgery in 2011, I have a new and profound understanding of behavioral health in regard to the body, mind, and spirit at the individual and social levels as never before. Body and mind are the easy constructs; spirit is a more difficult idea to apply.

Addressing the spirit

If the fire service wants behavioral health to help ensure everyone goes home, somebody needs to look at spirit as a way to significantly reduce occupational risks, injuries, and deaths.

Human beings learn the construct of the lie very early in life; we do it our entire lives, some better than others, and we may need the lie to be human. The lie is so common we may not know we are doing it.

One of the most powerful, funny, and disturbing movies ever is *The Invention of Lying*. It shows what society would be like if we did not have lying, how lying was accidentally discovered, and the power lying can have when used for good or bad.

Lying has its place from the Tooth Fairy to "You look good; did you lose weight?" Or in a not-so-good application from "I did not have sex with that woman" to "God told me to kill in his name."

Your next question is: "Clark, what the hell does this have to do with seatbelts?" Glad you asked.

On seatbelts

In 2012, nine firefighters and EMTs died in crashes without their seatbelts on.

NFPA 1500 requires all firefighters to be seated and belted when the apparatus moves. Every state governor and state legislator will tell you their firefighters' safety is important to them.

Yet, eighteen states exempt firefighters and EMS personnel from seatbelt laws—even the federal motor coach law exempts firefighters and EMTs when responding. If safety is important, why permit by law such unsafe behavior?

Can a state say that firefighter behavioral health is important if state law exempts them from using seatbelts? Is this some level of lie, even if unconscious or unintended? Once this issue is realized and not fixed, does the level of lie increase?

Tales from the street

In 2013 I had the privilege of attending a high-level meeting related to firefighter behavior and safety culture. All the major fire service organizations were represented.

The facilitator told a story of when he rode out with a metro fire department, explaining that seatbelts were not used. On a subsequent ride, there was no seat in the reserve apparatus, so the visitor rode on a milk crate. Seatbelts were again not used.

I then told the story of a West Point cadet who rode out with this same fire department; he too did not use the seatbelt all the time, just like the firefighters.

I asked my colleagues sitting around the table which organization will inform the fire chief of this department that the seatbelt policy is not being enforced and the fire department is not in compliance with national firefighter safety standards?

No one volunteered.

I struggled with this issue for a couple of weeks and asked friends for advice. Finally, I worked up the courage to write a personal letter to the fire chief and the three presidents of the employee labor organizations. It has been more than one month, and I have not received any replies to my letters.

All the major fire service organizations say that firefighter behavior health is important and spend millions of dollar on training programs, CDs, videos, and printed and online materials. All of these are important products that can lead to behavioral health.

But they could not write a letter.

Accepting the lie

If we see a wrong and do not try to fix it, is there some level of lie attacking the wellbeing of the body, mind, and spirit of our behavioral health quest? If there is a firefighter occupational death due to an unworn seatbelt in this fire department there will be a lot of finger pointing.

Saving our own is an inside job that starts in the firetruck. Captain Tucker Palmatier is my hero; the inside story below is in his own words.

How one fire captain changed his department's seatbelt culture
By Captain Tucker D. Palmatier, Glendale, Ohio, Fire Department

In 2010, the Glendale Fire Department held a department-wide training event where the National Fallen Firefighters Foundation presented its Life Safety Initiatives. During the training, the National Fire Service Seatbelt Pledge was explained and passed around for signatures.

Most of the firefighters and officers pledged to ensure all firefighters are belted in before leaving the station.

After a few months, the pledge started to fade from firefighters' memories. One afternoon, my crew was dispatched to a fire alarm drop.

As we left the station, I asked: "Does everyone have their seatbelts on?" An echoing "YES!" was given from the back seats. I looked at the driver and said, "Let's roll."

As we turned onto the main road, I noticed someone moving around in the back of the vehicle. I looked back to see one of the firefighters struggling to find his seatbelt. The seatbelt had fallen between the seats, and he was unable to locate it.

Knowing the answer, I asked again: "Does everyone have their seatbelts on?" Another firefighter responded, "Yes, we are good back here, just go."

I responded with, "Don't lie to me! I'm looking right at him" and ordered the driver to pull the truck over. My driver gave me a surprised look and pulled the engine over.

With the lights off and the siren winding down, we waited for the firefighter to get belted in.

The firefighters initially questioned my decision. I made it clear that I didn't care what the call was, if we were not safe, we were not going to proceed to the alarm.

Once the firefighter was buckled in, we resumed the response.

After the run, I notified the chief of my actions. I was anticipating the rumor mill to question my actions. I could already hear the "bumper talk" and grumbling in my head.

I was surprised when my chief was extremely supportive of my actions. He reinforced my decision and encouraged me to do it again if necessary. His support enabled me to do the right thing even though my safety concerns were not always popular with my fellow firefighters.

The catch phrase "Don't lie to me!" gained popularity among the firefighters and officers. It was often jokingly accompanied by, "Don't make me pull this engine over."

Whenever the officer in charge of an engine asked, "Is everyone belted in?" "Don't lie to me" became the sobering statement for everyone on the department.

It drove home the fact that firefighter safety was paramount. Merely saying you were wearing your seatbelt was not good enough.

Mandatory seatbelt use became so well supported that firefighters began buddy-checking each other. When the Glendale Fire Department purchased a new frontline engine, technology replaced the some of the guesswork involved.

Now a computer monitors the seats and seatbelt status. However, we continue to reinforce that it is never acceptable to remove your seatbelt while a vehicle is in motion.

Safety-conscious firefighters understand that if they don't get their personal protective equipment on before the run, they will wait until the truck stops at the fire scene. They will not get ready while the firetruck is moving.

Focusing on the seatbelt training provided by the NFFF may seem like a minor safety concern to some firefighters, but experience has demonstrated that it can save lives. Both leaders and firefighters alike need to insist that safety procedures are always followed.

At the end of the day, our most important mission is to ensure every firefighter returns home safely to his or her family.

Behavioral health

Seatbelts are just one example. Behavioral health is related to every behavior a firefighter does — from lighting a cigarette to taking SCBA off during overhaul, from going through a red light without stopping to running into a burning building without an attack line, from not reporting a seatbelt violation to receiving a medal of valor despite several safety SOP violations.

What level of lie does the fire service accept, consciously or unconsciously, about firefighter behavioral health and safety doctrine?

Captain Palmatier's actions represent the highest example of behavioral health courage and valor. I encourage some organization to give him an award for exemplary fire service wellbeing of the body, mind, and spirit. Or at the very least, write him a thank-you letter for getting us not to lie about seatbelts.

Telling the truth will increase the spirit of firefighter behavioral health and get us one click closer to "Everyone Goes Home." Ask this question for one of your drills during Safety and Health Week: "Do we tell the truth?"

Mayday! Mayday! Mayday!
Know When to Call It

Mayday! Mayday! Mayday!
Do Firefighters Know When to Call It?
2001

MAYDAY MAYDAY MAYDAY must be the most frightening three words that can be heard over the fireground radio. Everyone who hears the call knows that what was a public emergency that the fire department came to solve has now become an emergency for us. Something has gone wrong, and one of our own needs help.

Every fire department in the country has detailed SOPs explaining who on the fireground will do what when a firefighter calls Mayday. The RIT is activated, radio channels are changed, and additional chiefs and units are dispatched. We have all trained extensively on these procedures. We have developed special techniques on how to get downed firefighters out of tight spaces or up through holes. And we carry an RIT bag on the apparatus.

All this is important, but it is the easy part of the process. We have almost completely ignored the most important first step: Getting firefighters to recognize they are in trouble and need to get help, then to call Mayday.

I Can't Save You But I'll Die Trying: American Fire Culture

What Mayday decision parameters have we given firefighters? How do we teach the cognitive and affective Mayday decision-making process? How do we teach the psychomotor skill to execute the decision?

We have not answered these questions satisfactorily. Our standards and training are woefully lacking for this critical firefighter personal life saving competency.

The *NFPA 1001: Standards for Fire Fighter Professional Qualifications* (1997) does not definitively address the concept of Mayday. The word *Mayday* is not used in the standard. There is a Mayday inference in the Firefighter I Standard 3-2.3 that reads, "Transmit and receive via fire department radio." The firefighter is to know "Departmental radio procedures and etiquette for routine traffic, emergency traffic." The skill is "the ability to operate radio equipment and distinguish between routine and emergency traffic" (pp.1001–1).

Mayday is again alluded to in Standard 3-3.4. It reads, "Exit a hazardous area as a team"; knowledge "elements that create or indicate a hazard"; skill "evaluate area for hazard" (pp. 1001–7).

There is more verbiage on auto extraction than Mayday in the standard. In Firefighter II the only standard that comes close to Mayday is 4-2.3. It reads, "Communicate the need for team assistance"; knowledge "fire department radio communications procedures"; skill "the ability to operate fire department communications equipment." This standard seems to be about routine assistance, not Mayday conditions.

Chapter 23 of *The Firefighter's Handbook* (2000) has a section titled "Firefighter's Emergencies." The opening paragraph reads in part, "To help understand the actions to be taken during an actual or potential firefighter emergency, the firefighter must study procedures for rapid escape and declaring a Mayday for lost or trapped situations" (p. 690).

Under "Entrapments" it reads, "The first step a firefighter should take in an entrapment is to get assistance. Activation of a PASS device is warranted and the declaration of a Mayday should be made over the radio" (p. 692).

Under the heading of "Lost firefighter" it reads, "We cannot overemphasize that a fighter or team lost in an IDLH atmosphere is in fact experiencing a firefighter emergency" (p. 692). "First, the firefighter or team must report the fact they are lost. This is also a Mayday situation and should be transmitted as such over the radio" (p. 693).

Essentials of Fire Fighting (1998) does not refer to the word *Mayday*. In the "Rescue and Extrication" chapter a section is titled, "Trapped or Disoriented Firefighters." In reference to disoriented firefighters it states, "If they are not having any success finding their way out, they should find a place of relative safety and activate their PASS devices" (p. 181).

For trapped firefighters it states, "These firefighters should immediately activate their PASS devices. If either trapped or disoriented firefighters have radios, they should try to make radio contact as quickly as possible with other personnel on the emergency scene" (p. 182). Our Mayday standards and training doctrine clearly indicates that we have not researched the concept of a firefighter's calling Mayday scientifically.

To study the concept of people recognizing that they are in trouble and need help, I tried to do some benchmarking by looking to others who have addressed similar issues. The place I started with was Navy fighter pilots and the concept of ejection from their aircraft.

In terms of macho, firefighters and Navy pilots are about equal. This is the first assumption I made. Next, the decision to pull the ejection cord is similar to the firefighter's making the decision to call Mayday. Both the pilot and the firefighter are using their last resort to save their life. The ejection mechanism and our system to save downed firefighters are useless until the individual in trouble cognitively and effectively recognizes this fact and acts accordingly.

When the pilot punches out, the aircraft is lost. There is the potential for injury to people and property on the ground, and the pilot may be injured or killed. When a firefighter calls Mayday, other firefighters are put at risk to save him or her. The Mayday decision for the fire service must be considered extremely consequential.

The ejection doctrine for pilots begins as follows: "The first and absolutely most important factor in the ejection process is the decision to eject" ("Ejection seat training operations and maintains manual," p. 3-1, Environmental Tectonics Corp., Southampton PA 1999). "You should understand that the decision to eject or bailout must be made by the pilot on the ground before flying."

"You should establish firmly and clearly in your mind under which circumstances you will abandon the aircraft" ("Ejection seat trainer," p. 2. Environmental Tectonics Corp. Southampton, PA).

A key source of Navy ejection doctrine is the NATOPS manual for each aircraft. The Naval Air Training and Operating Procedures Standardization Program is a positive approach toward improving combat readiness and achieving a substantial reduction in the aircraft accident rate.

Standardization, based on professional knowledge and experience, provides the basis for development of an efficient and sound operational procedure. The standardization program is not planned to stifle individual initiative, but rather to aid the commanding officer in increasing his unit's combat potential without reducing his command prestige or responsibility. (Letter of Promulgation, W.D. Houser, Vice Admiral, USN, May 1, 1975).

The U.S. Navy F-4J jet fighter NATOPS flight manual (1995) contains the following ejection parameters:
- ■ If conditions for no-flap carrier landing are not optimum, eject.

- If neither engine can be restarted, eject.
- If a fire exists after catapult launch, should control be lost and not regained immediately, eject.
- If control speed/gross weight combinations exceed available arresting gear limits, eject.
- If field landing cannot be made, eject.
- If hydraulic pressure does not recover, eject.
- If carrier landing and all landing gear is up, eject.
- If carrier landing and one main plus nose gear up, eject.
- If the combination of weather, landing facilities and pilot experience is less than ideal, consideration should be given to a controlled ejection.
- It is recommended that a landing on unprepared terrain not be attempted with this airplane, the crew should eject.
- If still out of control by 10,000 feet above terrain, eject.
- If the flap and or BLS failure occurs during the catapult stroke or shortly thereafter, eject immediately.

It is important to remember that each different type of aircraft has its own ejection parameters. Pilot ejection training consists of classroom and flight simulator to develop cognitive and effective skill. Then the ejection seat trainer is used to imprint the psychomotor skill. Ejection retraining occurs every six to twelve months.

The failure or delay to eject can be attributed to ten reasons that must be addressed in ejection training, according to Richard Leland, director of Aeromedical Training Institute Environmental Tectonics Corp.

1. Temporal Distortion (time speeds up/slows down)
2. Reluctance to relinquish control of one's situation

3. Channeled attention (i.e., continuing with a previously selected course of action because other more significant information is not perceived)
4. Loss of situational awareness (i.e., controlled flight into terrain)
5. Fear of the unknown (i.e., reluctance to leave the security of the cockpit)
6. Fear of retribution (for losing the aircraft)
7. Lack of procedural knowledge
8. Attempting to fix the problem
9. Pride (ego)
10. Denial (i.e., This isn't happening to me.)

By now some readers are thinking fighter pilots have it easy because the instruments in the cockpit do not change. The positions of the needles move and when enough gauges are in the red it is time to eject.

Firefighters do not have gauges to read or clearly defined input data, and the critical information is dynamic throughout the emergency event. Each type of structure we enter—single family, duplex, garden apartment, triple-decker, high-rise, commercial, industrial, taxpayer—may require specific Mayday decision parameters. Once we determine the parameters, we need to recognize them and act correctly. Will we?

In 2000 the Chesterfield, Virginia, Fire Department conducted a lieutenant's test. Part of the testing included a field activity. Seventeen candidates for lieutenant were taken to a large abandoned building, eighty by one-twenty feet with an open floor plan. One at a time, in full turnouts, SCBA with less the seven hundred PSI, portable radio, and Nomex hood on backward covering their facemasks, each candidate was taken into the building and told: "You are the OIC of the first engine operating at a fire in a Shopping Mall.

"You and your crew are stretching a 1¾" hand line at the top of the escalator on the second floor and you encounter 'cold' smoke and zero visibility. While maintaining voice contact with your crew, you have been searching for the fire. You no longer have voice contact with your crew and are now lost and disoriented. This is not a training scenario; your life depends on your actions!" ("Test asks: Can you Survive?" Heather Casey, Firehouse.com News, Sept. 28, 2000) The correct actions to take were:

- Declare an emergency on the radio.
- Activate the emergency button.
- Announce "Mayday, Mayday, Mayday, Emergency Traffic."
- Activate the PASS device.
- Successfully merge with the RIT.

Of the seventeen candidates, only four took the correct action immediately. The fastest times to complete the tasks were four to five minutes. Some of the candidates never called Mayday (Personal communications, Captain Dave Daniels, Chesterfield Fire Department, Sept. 25, 2001).

This outcome should raise concern for all of us because the candidates were put into the Mayday decision parameters and most did not make the correct decision immediately. In other words, they were told the gauges were in the red and still did not react correctly.

Remember, on the real fireground each firefighter must read the gauges, determine the meaning, and then make the Mayday decision.

Again I ask. What are the Mayday decision parameters for firefighters? How do we teach the Mayday decision-making process to firefighters? How much Mayday practice do firefighters need?

I don't have the answers to these questions. The military aviation method of creating ejection doctrine may serve as a model for us to use in answering these questions. We need to get our best minds researching the questions to create firefighter Mayday doctrine.

I do know this. A firefighter's decision to declare a Mayday is made in the fire station before getting on the apparatus. So, at your next company drill, ask this question: When would you call Mayday?

When Would You Call Mayday! Mayday! Mayday?
2002

I HOPE YOU WILL never need to call Mayday for yourself or any other firefighter. But you need to be prepared to do so because your life may depend on this single decision.

When firefighters are asked, "When would you call Mayday?" you get some unexpected answers like: "I push the orange button on my radio" or "I don't have to worry about that because I am on the engine company and I have the hose line to find my way out." It is the truckees that go above the fire that need to call Mayday." These are actual answers from career firefighters in large metro fire departments.

When you push firefighters to answer the question they will usually rely on the statements in their SOP like, "When Lost-Missing-Trapped and their life is in danger firefighters will announce Mayday-Mayday-Mayday." When you ask firefighters to give an example of lost, missing, or trapped they have a difficult time coming up with a specific example. Then they start including statements like, "It depends on your experience" even though they have never had the experience of calling Mayday.

The problem is that we have not clearly defined *lost*, *missing*, or *trapped*. We leave it up to each firefighter to define these terms. Somehow we think firefighters will intuitively know when to call for help. This is a very dangerous assumption. Currently we do not teach firefighters when and how to call Mayday at the cognitive, affective, and psychomotor levels of learning to the mastery level of performance.

If firefighters must perform a decision-making process and execute a set of skills very rarely or never in their careers but the decision and behavior have life-or-death consequences, they must be trained and retrained throughout their careers.

We developed a draft Mayday Decision Parameters for a Single Family Dwelling. The SFD was selected because it is a basic type of structure fire common to many fire departments, it is a high risk to firefighters, and it was describable. Keep in mind that we will need a Mayday Decision Parameter for each type of structure we enter. A qualitative method was used that included brainstorming (individual and small group) to create the specific parameters. (The first research team to help develop these parameters were John Koike, Dennis Culbertson, Tommy Harmon, Linda Pellegrini, and Tom Wiley of the NFA Interpersonal Dynamics Class, December 20, 2001, instructors Paul Burkhart and Howard Cross, research advisor Burton Clark). An opinion survey, using convenience sample populations (N=339), was used to determine if firefighters agreed or disagreed that they must call a Mayday under specific conditions.

This methodology has significant limitations because it relies on judgment and opinion. The results are not conclusive and have not been field-tested. They are presented only to foster further discussion and study of fire service Mayday doctrine.

Survey Results: 339 Respondents

Mayday decision parameters: single family dwelling detached, one- or two-story with or without basement* IDLH environment/SCBA in use

Firefighters must call a Mayday for themselves under these conditions.

% said YES	Possible Mayday Conditions
98%	Tangled, pinned, or stuck; low- air-alarm activation: Mayday
94%	Fall through roof: Mayday
92%	Tangled, pinned, or stuck and do not extricate self in sixty seconds: Mayday
89%	Caught in flashover: Mayday
88%	Fall through floor: Mayday
82%	Zero visibility, no contact with hose or lifeline, do not know direction to exit: Mayday
69%	Primary exit blocked by fire or collapse, not at secondary exit in thirty seconds: Mayday
69%	Low air alarm activation, not at exit (door or window) in thirty seconds: Mayday
58%	Cannot find exit (door or window) in sixty seconds: Mayday

*Assumptions: SFDs usually have a front door and back door. Most rooms, except for bathrooms, have at least one window that could be used as an exit. The exception to door and window assumptions will be the basement, attic, hallways, closets, storage areas, and attached garage. NOTE: SFDs with barred windows or windows too small or too high from floor to use as an exit are excluded from this MDP.

Respondents: This was a convenience sample made up of National Fire Academy students (N=181), Executive Fire Office Program graduates (N=96), and Fire Department Instructors Conference students (N=62); all respondents read the original Mayday article and or were given an oral briefing on its contents before answering the survey. The responders ranged from recruit firefighters to fire chiefs, career and volunteer, small rural to large metro.

A significant challenge to firefighters under IDLH conditions is carbon monoxide affecting their judgment, motor skills, and sensory perception. In addition the environmental conditions—smoke, heat, gases, and structural stability—can quickly become deadly. The rapid intervention team takes time to rescue a firefighter; the window of survivability can be small.

The same ten factors that cause pilots to fail or delay ejection may apply to firefighters' failing or delaying to call Mayday. Is it better for one hundred firefighters to call Mayday and not need it, than one firefighter not to call Mayday and need it? By reacting to decision parameters, a firefighter's perceived need for help is eliminated from the decision-making process. For example, if you fall through a floor, you may not be injured; there may be no fire or smoke; you may be able to get up and walk right out of the building. The condition of falling through the floor is not normal—something has gone wrong, your judgment is impacted, and the event may be fatal. Calling Mayday immediately is the only one-hundred-percent correct response and that still does not ensure survivability. The fire service has rules to protect us: Wear your seatbelt, stop at red lights, wear your SCBA, use BSI, have a backup spotter. We do not rely on the firefighter's perceived need to comply with the rule or experience of the consequences to comply with the rule. Firefighters are expected to follow the rules, and we hold them accountable. No one gets in trouble for following the rules.

What are the rules for calling Mayday? The purpose of this article is to generate discussion and research on fire-service Mayday doctrine. The questions we need to answer are: What are the Mayday decision parameters for firefighters? How do we teach the Mayday decision-making process to firefighters? How much Mayday practice do firefighters need? When would you call Mayday? That is a good question to ask all the firefighters in your department. Let us know if they all get the answer one hundred percent correct.

Captain Steven Auch, Indianapolis Fire Department, and Captain Raul Angulo, Seattle Fire Department, contributed their knowledge and expertise to this article.

You Must Call Mayday for RIT to Work. Will You?

2003

YOU HAVE PROBABLY participated in some type of rapid intervention team or "Saving Our Own" training, and your SOPs may have some directions on a Mayday. The odds are, however, that you have not been given specific rules on when to call a Mayday. You are taught to be the rescuer, not the victim, and your recognition-primed decision-making process (defined below) may interfere with your calling a Mayday when you should.

What does this mean for firefighters? First, it means that we've put the cart before the horse. It doesn't matter how well-trained or well-equipped your RIT is. Unless the incident is witnessed, RIT teams won't be activated unless you or your partner calls a Mayday. The training emphasis has been on saving our own, not on our own calling for help. We would hate to speculate, but some firefighters might have survived had they recognized early enough that they needed help or that something was out of the norm and called a Mayday.

Deputy Assistant Chief Curt Varone of Providence, Rhode Island, has verified our thoughts by identifying eleven structure fires between 1978 and 2002 in which failing to call or delaying a call of a Mayday contributed to twenty-four line-of-duty deaths.[1]

Firefighters do not like to admit that they might need to be rescued. The delay in calling a Mayday may be caused by many factors, but three need to be addressed immediately: 1.) the stigma associated with admitting to yourself and letting others know you need help; 2.) not having been given clear rules for calling a Mayday; and 3.) the manner in which the fire service makes decisions. Last year, the Seattle, Washington, Fire Department had three near-miss incidents involving firefighters in interior firefighting operations. Each of these incidents easily could have led to LODDs, had help taken a few more seconds to arrive. The particulars of these incidents were detailed in "Train in 'the Rule of Air Management'" (*Fire Engineering*, April 2003). All three firefighters—a captain, a lieutenant, and a firefighter—are seasoned veterans and well-respected members of the department.

There were some disturbing similarities in the three incidents:

- None of the firefighters in distress called for a Mayday.
- None of their partners called for a Mayday.
- No one activated the emergency button on the radio.
- No one activated the PASS device.
- None of the partners activated a PASS device.
- Each firefighter became separated from his partner.
- Each firefighter ran out of air.
- Each firefighter suffered debilitating effects of carbon monoxide.

When interviewed, one firefighter said, "I knew I was in trouble. I thought about using the radio, but I thought, I found my way in; I can find my way out." Peer pressure and the "stigma" surrounding the idea that help is needed played a part in each incident. These firefighters realized that events were not unfolding correctly. They were all trying to find their way out of the building, but they couldn't. They all ran out of air. They all tried alternate filter-breathing techniques. But in the end, exposure to carbon monoxide impaired their judgment and motor skills.

Establish Mayday decision-making parameters

To ensure that firefighters will call for help as soon as they recognize that they may be in trouble, fire departments need to develop clear Mayday decision-making parameters (rules that specify when a Mayday must be called) and institute Mayday training programs firefighters must take and continue to pass throughout their fire-service experience. The parameters/recommendations are based on logic similar to that used to establish training programs that teach military fighter pilots when they should eject from their planes in an emergency.[2]

Fighter pilots are given clear, specific ejection parameters (rules governing when to eject), and they are trained and retrained on making the ejection decision and drilled on actually pulling the ejection cord several times a year. The comparison of firefighters calling a Mayday to pilots ejecting from their planes makes good sense, according to Kelly M. Woods, a former Navy fighter pilot who had to eject over North Vietnam when his jet was shot out from under him. After military service, he became a career firefighter. He and his partner were advancing a line down a basement stairway when the stairway collapsed, pinning him under the stairs. His partner called a Mayday. Today, Woods is an instructor with the West Virginia State Fire Academy.

It may seem strange that we have to create rules to tell firefighters to call a Mayday. But remember that we teach firefighters to be aggressive and expect them to act aggressively. Chief Alan Brunacini of the Phoenix, Arizona, Fire Department noted at the 2002 Maryland Fire Chiefs Conference:

"The hardest thing to do is to put a firefighter in reverse." Think of how we train firefighters. Do they ever fail to put out the fires in rookie school, or do they ever have to make the decision to retreat? Are firefighters ever put into training or drill situations in which they have to make the decision to call a Mayday for themselves? If the answer to these questions is *no*, how can we expect our firefighters to make these decisions under real-world life-and-death conditions?

The decision-making method

The manner in which we make decisions may be part of the problem also. Klein Associates researchers analyzed how U.S. Army battlefield commanders make decisions. We are using the military/fire service comparison because firefighters, like the military, must make decisions "while confronting time pressure, [under] changing conditions, [for] high stakes, and [with] unclear immediate goals and incomplete information."

The Klein study describes the cognitive process used to make decisions on the fireground, referred to as "Recognition-Primed Decision-Making (RPD)." As an example, officers arriving on the scene look at the picture (visual cues: fire, smoke, construction, time of day, occupancy, and so on) in front of them and then compare that picture with the pictures in their memory banks. When a match is found, they choose what worked at a similar situation in the past and use that experience to drive their strategy and tactics for the current situation.

This is a very rapid decision-making process. The first option chosen and followed is also most likely the only option considered. RPD is effective most of the time but not all of the time. Kline states: "Unfortunately, the first option may not be the best decision." This memory bank of pictures and actions we have to choose from has been developed over years of experience and training. It has been referred to as a "photographic slide tray." Using this analogy, we might say that "we may be missing some slides." RPD isn't limited to command-level officers; we all use it.[4]

RPD and Mayday

What does RPD have to do with Mayday? Remember that all three Seattle firefighters, two officers and one firefighter, were experienced. They had gotten themselves out of tight spots before; all said they had experienced running out of air and using the filter-breathing method (disconnecting the low-pressure hose from the regulator and putting the end in the turnout coat to breathe) to get out at previous fires. None had ever had a Mayday called for them. They were using RPD to respond to the situation at hand, but it did not work this time. It is safe to assume that the Mayday-calling slide was not in their RPD slide tray.

Do you train firefighters in the simple act of using the radio to practice calling a Mayday? If not, maybe you should. For example, at a working fire, an officer fell through the floor into the basement. His radio transmission was, "14's in the basement."[5] He never called a Mayday. Other factors also contributed to this LODD. We do not know if he had the Mayday-calling slide in his RPD slide tray.

Our firefighters may not be prepared to call a Mayday for themselves. Following is a summary of research conducted for previous articles. The tests covered making decisions pertaining to calling a Mayday.

■ The New Iberia, Louisiana, Fire Department conducted a drill to determine if the firefighters would call a Mayday for themselves. An open-space sixty-by-one-hundred-foot building was used; four hundred feet of hose was stretched through the building, and eighteen teams of two members and one team of three members were sent in one team at a time. They were told to follow the hose and assist another team at the end of the hoseline. The conditions were immediately dangerous to life and death (IDLH), cold smoke, and zero visibility (masks were blacked out). Their SCBAs had only eight hundred PSI in them. (Only three firefighters noted the low air.) Thirty-nine members participated—seventeen captains, fourteen drivers, and eight firefighters. All personnel had a portable radio assigned to them on the apparatus; only eighteen of the thirty-nine firefighters took their radios in with them. The situation made it impossible to fulfill the assignment of joining the other team at the end of the hose.

Training Officer Martin Delaune reported the following:
- Four kept going until their air was depleted.
- After the low-air alarm activated, twenty-two kept going forward for four minutes.
- After the low-air alarm activated, eight kept going forward for three minutes.
- Two discussed the situation for 2½ minutes before beginning the retreat after alarm activation.
- Three began the retreat when the low-air alarm activated.
- Three activated their PASS alarm.
- Two radioed a Mayday.
- None survived. They ran out of air before they got out.[6]

■ The Fort Worth, Texas, Fire Department tested about five hundred firefighters (four companies at a time) in a RIT/Mayday drill. A large open-floor plan building was used. A charged 145-foot 1¾-inch attack line went from the entrance door into the building. One loop had been placed in the line. The conditions were IDLH and zero visibility (masks blacked out). The line ended at a doorway that led into a suite of three offices. A manikin was placed in one of the rooms. The teams were told to rescue the downed firefighter near the nozzle. About one-quarter (130) of the firefighters were unsuccessful in exiting the building before they ran out of air. Most did not call a Mayday; all were declared non-survivors. The few that called a Mayday for themselves made the call outside the window of survivability.

■ The Indianapolis, Indiana, Fire Department used a 2½-story wood-frame residence charged with live smoke for department-wide RIT training. Four-member RIT teams were activated to locate a trapped firefighter who had declared a Mayday. Department Training Chief Doug Abernathy estimates there were fifteen to twenty failures of the low-air warning system on the SCBAs worn by the rescuers. Many of the failures resulted in out-of-air situations. Other firefighters became separated from their partners. None of the rescuers called a Mayday for themselves. "We found that we have a long way to go with our RIT and Mayday training," Abernathy reported. Washington Township, a department adjacent to Indianapolis, recently tested 120 firefighters in a Mayday situation. Using a large, recently abandoned restaurant and blacked-out facepieces on the SCBAs, the firefighters were taken in one at a time. With the low-air warning already sounding, all were told that they were members of the attack crew. It was further explained that they had become separated from the others. Individually, the firefighters were spun around, to disorient them, and positioned five feet from the charged handline. Training Officer Dale Strain explained that he hoped the firefighters would then declare a Mayday over the radio and activate the alarm on their PASS device. Strain reports that all but a few did one or both procedures; he attributed this success to the Mayday training the firefighters had recently received.

Mayday rules

Firefighters start developing their RPD slide tray in rookie school. Hesitation, retreat, and call for help are not learned. With this in mind, how do we learn when to call a Mayday? Throughout your career you will most likely never need to call a Mayday. We cannot rely on experience to teach us this competency—the first time may be the last time. If there is a very important skill that you very rarely need to use and you have to do it right the first time, you must drill, drill, drill—drill your entire career. Jetfighter pilots review ejection doctrine before each takeoff, and they drill on it every two months.

We developed nine "Mayday Decision Parameters" to guide firefighters in deciding when to call a Mayday in a single-family dwelling fire.[7] Individuals and small groups brainstormed to identify the specific parameters. The parameters were then submitted to sample populations of firefighters (339), to determine if they agreed or disagreed that they must call a Mayday under those conditions. These parameters are not conclusive and have not been field-tested. The nine conditions receiving the highest number of "agreements" among those surveyed that these conditions warrant calling a Mayday are presented to foster further discussion and study.

The parameters are as follows:
1. If you become tangled, pinned, or stuck and the low-air alarm activates.
2. If you fall through the roof.
3. If you become tangled, pinned, or stuck and do not extricate yourself in sixty seconds.
4. If you are caught in a flashover.
5. If you fall through the floor.
6. If there is zero visibility and no contact with the hose or lifeline and you do not know in which direction the exit is.
7. If your primary exit is blocked by fire or collapse and you are not at the secondary exit in thirty seconds.
8. If your low-air alarm is activated and you are not at an exit door or window in thirty seconds.
9. If you cannot find the exit door or window in sixty seconds.

It would seem that firefighters intuitively would call a Mayday if they fell through the floor. However, when we asked 339 firefighters from many different fire departments if they would call a Mayday if they fell through the floor at a single-family dwelling fire under IDLH conditions, only eighty-eight percent said they would. What are the other twelve percent going to do? Whatever it is, it is not the correct first decision. Ninety-eight percent said they would call a Mayday if they were tangled, pinned, or stuck and their low-air alarm activated. That still leaves two percent who would not call a Mayday.

The Mayday condition with the lowest "yes" response was, "Cannot find exit (door or window) in sixty seconds." Fifty-eight percent said they would call a Mayday; forty-two percent said they would not. Remember, this fire example was in a single-family dwelling—front door, back door, and window in most rooms. We did not choose this dwelling or the exit Mayday condition by accident. When you review the NIOSH firefighter fatality reports for one- and two-family dwellings, the firefighter victims were very close to a window or exit door but still failed to get out in time. One minute (sixty seconds) can be an eternity. Managing air and time in IDLH conditions are critical factors in Mayday decision-making.

Recommendations

We encourage you to be creative and to address these issues by yourself, with your crew, with your department, and with your trainees and to implement training programs that incorporate these conditions and procedures for overcoming them. Practice calling a Mayday over the radio. Blindfold the firefighters. Have them wear gloves; hand them the radio and see if they can turn it on, get the correct channel, push the emergency identifier button, push the talk button, and verbally call a Mayday. Have someone on another portable radio serve as communications and receive the information: Who is calling? What is the problem? Where do you think you are? Repeat the same drill in full turnout gear with SCBA in use.

Put some mattresses on top of the firefighters. See if they can get the radio out of their pocket.

As the company officer, tell your crew when you expect them to call Mayday for themselves. Give specific examples. Tell them when you will call a Mayday for them, giving specific examples such as under IDLH conditions or "if your leg falls through the floor and I cannot pull you out on the first try, I will call a Mayday" or "if the ceiling falls on us and we get tangled in wire, we will call a Mayday and then start cutting our way out.

At the training academy, every time you have live-fire training, place crew members in a situation in which they must make the Mayday decision for themselves. The instructor can drop a cargo net over a member or block the exit. Build a prop that drops the firefighter through a trapdoor into a ball pit. This will also create a drill in two-in/two-out and RIT. It will also desensitize the others on the operational team to the Mayday call so they continue fighting the fire instead of abandoning their assignment to go to the aid of the downed firefighter.

If we want RIT and Saving Our Own to work, we need to put the Mayday calling slide into every firefighters RPD slide tray. Then, we need to drill on it often. Because RPD "is predicated on people choosing a course of action based on pattern matching, a comparison of the current problem to similar problems encountered before." We cannot rely on fireground experience to teach us when to call a Mayday; therefore, we must simulate this lifesaving skill often.

A sobering thought related to the issue of RIT and Mayday comes from Battalion Chief Kenny Freeman of the Fort Worth, Texas, Fire Department: "Personally, perhaps the most important issue brought to light through the RIT training involves the realization that my expectations and assumptions concerning the deployment of an RIT team were both inaccurate and unrealistic. While my previous assumptions were totally born out of a commonly held perspective, they would have been nonetheless ineffective and possibly tragic in the final analysis.[8]

Rapid Intervention Teams and Saving Our Own training are wonderful firefighter survival tools. But, like all safety equipment or SOPs, the most important component is the firefighters themselves. Just as you have to put on your seatbelt to have it protect you in an accident, you have to call a Mayday for the RIT to come to get you out. Will you?

Captain Steven Auch, Indianapolis Fire Department, and Captain Raul Angulo, Seattle Fire Department, contributed their knowledge and expertise to this article.

References

1. Varone, C., "Firefighter safety and radio communications," *Fire Engineering*, March 2003, 141-164.
2. Clark, B. "Mayday, Mayday, Mayday: Do firefighters know when to call it?" Firehouse.com, October 2001.
3. Burkell, C. and H. Wood, "Make the right call," *Fire Chief*, March 1999, 42.
4. Varone, C., "Not your fathers command post," *Fire Chief*, August 2001, 72.
5. "Report from the reconstruction committee fire at 400 Kennedy Street, NW," District of Columbia Fire Department, Washington, DC, October 24, 1997.
6. Personal communication, March 9, 2003.
7. Clark, B., S. Auch, and R. Angulo, "When would you call Mayday Mayday Mayday?" Firehouse.com, July 2002.
8. Kenny Freeman, personal communication, March 26, 2003.

Calling A Mayday: The Drill
2004

THANKS TO THE cooperation of the Anne Arundel County Fire Department, the Maryland Fire Rescue Institute, and the Laurel Volunteer Fire Department, the firefighter Mayday concepts presented by Clark (2001, 2003) and Clark, Auch, & Angulo (2002, 2003) were put to the test and passed with high marks. The Mayday Doctrine theory is based on an analysis of the engineering, psychology, physiology, and training aspects of a firefighter calling a Mayday. This analysis used jet fighter pilot ejection doctrine models as the foundation (benchmark) for developing firefighter Mayday Doctrine.

Over a three-day period ninety-one firefighters and officers experienced what it might be like to call a Mayday using their cognitive, affective, and psychomotor skills. The overwhelming conclusion by all who participated was that everyone needs this type of training and it needs to be repeated throughout time in the service. Battalion Chief Dave Berry of the Anne Arundel County Fire Department conducted the training for Battalion 3 on all three shifts. The drill consisted of classroom lecture and hands-on practice. Each class size was about fifteen students, with two drills per day (morning and evening) for six drill deliveries total.

Chief Berry used the Mayday articles as the foundation for the lecture portion of the battalion drill, "Calling a Mayday." In addition he asked 110 firefighters "What Makes You Call a Mayday?" From this extensive list he narrowed the Mayday parameters down to six words: fall, collapse, activated (low air or PASS device), caught, lost, trapped. To drive home the need for Mayday training, the Seattle, Washington, Fire Department videotape of the three firefighter near-misses was presented. This tape clearly illustrates how quickly a firefighter becomes incapable of calling the Mayday because of carbon monoxide, which reduces cognitive decision-making and small motor skills, and the psychological reluctance of firefighters to call for help. An additional videotape of the near-LODD of an Anne Arundel County firefighter brought the point home that this can happen to you and you only get one chance to call Mayday correctly.

The most elaborate prop simulated falling through the floor. This prop was designed and built by engineering technician Donny Boyd of the MFRI. The prop consists of a ramp the firefighter crawls up. At the top is a teeter board, which tilts forward when the firefighter crosses the center of gravity, dumping the firefighter into the third part of the prop, the ball pit. The ball pit is actually filled with cut-up swim noodles because they were less expensive than balls and are more durable. A key concern was safety of the firefighter. No one was hurt, but the firefighters knew that they had suddenly fallen into something. The transportable prop was build for under one thousand dollars.

The second prop, simulating a ceiling collapse, was made of chain-link fencing that was dropped over the firefighters as they crawled under it. Two instructors then stood on the fence, restricting the firefighters' movement and making it impossible for them to escape.

The classroom lecture also covered the three AACFD procedures for calling a Mayday. First, push the emergency identifier button on the radio. This captures the channel for twenty seconds, gives an open mike to the radio (in other words the firefighter does not need to push the talk button on the radio), and sends an emergency signal to radio communications identifying the radio. Second, announce Mayday, Mayday, Mayday. Third give LUNAR: **L**ocation, **U**nit number, **N**ame, **A**ssignment (what were you doing?), **R**esources (what do you need?). The classroom portion of the drill took about ninety minutes. Chief Berry distributed a job aid the size of a business card to all participants; it listed the six Mayday parameters on one side and the three procedures for calling a Mayday on the other side.

The hands-on portion of the drill took place in the basement of the fire station. The Mayday props were set up before the drill, and the area was placed off limits so no one knew what they were to experience. The four Mayday props simulated falling through a floor, being pinned under a ceiling collapse, getting lost/trapped in a room, and becoming stuck while exiting the structure.

The third prop was a small bathroom with a sink and toilet about five by six feet. A hose line with nozzle ended in this room. Once firefighters were inside, the door was closed and a wooden chock placed under the door. This made it impossible to exit the room.

The fourth prop simulated becoming stuck while exiting a building. A piece of wire rope with a slip loop was dropped over the firefighter's SCBA bottle. As they continued crawling the loop tightened, making it impossible for them to move forward. Try as they may, they could not get loose.

One at a time the firefighters were brought to the outside basement entrance. They were in full turnout gear with SCBA. At the entry point they were given the assignment. "This is a simulated fire with IDLH conditions. You and an imaginary partner are to follow this attack line into the kitchen. When you arrive your assignment is ventilation." The firefighters were reminded of LUNAR, put on air and their facepiece blacked out. The door was opened. They were told to go on hands and knees and follow the hose line.

The firefighters immediately had to crawl up the ramp (spotters were on either side); when the teeterboard tilted, they fell into the ball pit. The firefighters were expected to call a Mayday. If that was not their first reaction, the instructor prompted them: "What just happened to you?"

Answer required: "I fell into something."

Prompt: "What are you to do if you fall?"

Answer required: "Call a Mayday."

Prompt: "Correct, do it."

After the firefighters correctly pushed the EIB, said Mayday Mayday Mayday, and gave LUNAR they were told that they were done and were helped out of the ball pit. The instructor then reset the radio. They were told to go down on hands and knees again, crawl to another line, and continue their assignment. After crawling about fifteen feet, the chain link fence was dropped on them. The instructors stood on the fence making it impossible to escape. The correct response was to call a Mayday. If the firefighters struggled for more than a minute, they were prompted again. After calling the Mayday, they were released, their radio was reset, and they were told to continue their assignment. After another fifteen-foot crawl, they ended up in the bathroom at the nozzle; the door was chocked closed. This put them in the lost or trapped Mayday parameter. If after two minutes of trying to get out they did not call a Mayday, they were prompted. After the correct response, they were let out of the bathroom and the radio was reset. Next, they were told to find a nozzle on the floor outside the room they had just left, then exit the building by following the line. The line took them around a metal fence/guardrail to a wheelchair ramp that led to the exit.

As they turned the corner, a wire rope was dropped over the firefighter's SCBA bottle without his or her knowledge. After crawling six feet, the rope tightened, and they were stuck. After a minute of trying to get loose, if they had not started to call the Mayday, they were prompted.

Lessons learned

At the first prop, most all the firefighters had to be prompted to call the Mayday. Their first instinct was to get out of what they had fallen into. The instructors did not let them get out. Their next challenge was pushing the EIB. This proved to be difficult for most of them and caused frustration and anxiety. The anxiety was evident by the increase in their breathing rate. The frustration was evident when some tried to remove a glove to find the button. Instructors did not allow this. They were prompted, "You just burned your hand. Put the glove back on." Most tried reaching down into the pocket to activate the EIB, which usually proved unsuccessful. Some had to take the radio out of the radio pocket; in many cases this manipulation of the top of the radio caused them to change the radio channel. The longest time to successfully push the EIB was two minutes. Because of the frustration and anxiety, the LUNAR report was not always given correctly. The frustration and anxiety were most likely because this seemingly simple skill of pushing the EIB was not easy. Pushing the emergency identifier button was challenging because the radio sat too far down in the radio pocket, gloved hands made it very difficult to activate the EIB, and the radio was a style new to the department.

At the second prop, the firefighters quickly realized they were not getting out of whatever had fallen on them, so few needed to be prompted to call the Mayday. This time restricted movement challenged them because the fence was all around them. Many had to remove the radio from their pocket. Since they had performed the EIB skill once before, they knew they could do it, so they just kept working at it. As the firefighter's EIB skill proficiency level increased, the LUNAR transmission was more accurate.

At the third prop there was no restriction on them physically. Many tried to break down the door; they were not allowed to do that. Most still had to remove the radio to activate the EIB. They gave LUNAR, but few reported that they were in a bathroom. Only one needed to be prompted to call the Mayday after about two minutes of just sitting in the room.

At the fourth prop, they were tired and quickly realized their forward movement was stopped. In most cases the "swim technique" did not reveal the rope, so they called a Mayday. Their LUNAR usually did not include the fact that they were now trying to exit the building they were still reporting "division one, kitchen, ventilation, trapped."

Only one firefighter was observed to have no difficulty pushing the EIB in the pocket; he even did it without lifting the pocket flap. During the second drill period, Firefighter J.B. Hovatter was observed not putting his radio down in the pocket. He had taught himself to put the pocket flap down inside the pocket and hook the radio clip over the chest strap of the SCBA. This technique positioned the radio halfway down in the pocket, keeping the controls outside the pocket but still securing the radio to the firefighter. He quickly activated the EIB every time. It was decided to teach this technique, "The Hovatter Method," to all remaining firefighters, whose performance level increased dramatically.

A discussion session was held with the class after each drill to show what the props were and to get feedback. Overwhelmingly, they said it was an important learning experience, and they all agreed the drill should go department wide.

Feedback

Division Chief Allen Williams, health and safety officer for the AACOFD who observed the drills, said: "Hopefully firefighters will do all they can to not need to call a Mayday. However, firefighting is dangerous and the risk is there. Firefighters are reluctant to call Mayday. The training forced them to call Mayday. The training was excellent. The training is a very good risk-management strategy."

Battalion Chief Dave Berry said: "This training shocked them into calling a Mayday. It took some of the bravado out of them. It doesn't matter what rank you are, we can all get into a situation where we need to call Mayday. The drill became the great equalizer. In training it is difficult to shock a person into calling Mayday without hurting them; these props can do that. I know now that my battalion can call a Mayday if they have to."

Captain Leroux said: "We needed to be coached through calling a Mayday; it did not come naturally. We had machismo and self-doubt. Should I or shouldn't I call Mayday; I'll be embarrassed. We learned how important it is to call Mayday quickly while you still can think and explain where you are and answer questions. It is my crew and I that go in and will be using this skill. When you get in a Mayday situation you are going to be so stressed out—calling Mayday has to come natural and this training will help."

A firefighter: "When they dropped that fence on me I realized I was done. You are calling people to come get you out. I had to concentrate on getting to the button and calling a Mayday."

Some veteran firefighters said, "It was the best training we have ever received in our career."

Findings

For the Mayday call to be completed it must be received by someone in communications, then communications must repeat back to the firefighter the information reported. This is the only way the person calling the Mayday will know the message was received correctly.

The hands-free feature of the radio is useful, but if the mike is turned facing the firefighter's coat the message will become muffled.

The firefighter must speak loudly, clearly, and distinctly to be heard and understood.

If LUNAR is not the normal day-to-day communications sequence for talking on the radio it may not come naturally to firefighters under Mayday conditions.

In some cases the radio EIB did not reset correctly. The next time the EIB was pushed the three beeps sounded indicating the open mike was on but there was no transmission.

It was learned that AACOFD communications could reactivate the captured channel and open the mike for an additional twenty seconds and repeat opening it as needed.

The AACOFD is working on purchasing user-friendly firefighting gloves. This will help in using the radio.

Situational awareness can be compromised very quickly in a zero-visibility environment.

The fact that you decided to call a Mayday can tax your higher cognitive thinking, like where you are and what you are doing, which are important facts for the RIC.

Calling a Mayday is a complicated cognitive, affective, and psychomotor skill set that relies on a radio and the communication system, both human and hardware, that gets the call for help. A failure in any component part of this system can be disastrous. We need to study, test, train, and drill the entire Mayday calling system if we expect it to work when we need it.

Recommendations

First, practice calling Mayday. Can you push the EIB in five seconds with all your gear on? What happens when you push the EIB? (Does the radio channel change, who receives the EIB signal, where is it received, what do they do with the information?) Can you get to the radio when you are covered with debris? Where does the mike need to be so you can be heard? How loudly do you need to talk?

Second, include Mayday calling as a subset drill in all training where firefighters are put into simulated IDLH conditions. At a minimum, in rookie school and throughout their service, firefighters need to practice calling Mayday as often, if not more than they practice tying knots. Our bodies and minds need to be shocked into Mayday parameters repeatedly so the correct response becomes natural and instantaneous.

Third, get communications involved. How many times do dispatchers practice receiving and responding to a Mayday call? You do not want your real Mayday call to be the first time the radio operator gets to test their Mayday skills, radio equipment EIB function, and Mayday procedures.

Finally, whether you are the rookie firefighter or fire chief, if you put on SCBA and enter IDLH environments, you need to drill on "Calling a Mayday."

Author's note

After the pilot delivery of the drill in Battalion 6, the department moved the class to the county fire-training academy. Chief Berry was assigned to conduct the drill for the entire department. As of the end of June 2004, all seven hundred career and three hundred active volunteer personnel in the Anne Arundel County Fire Department had gone through this "Calling a Mayday Drill." Congratulations to the first fire department in the nation to do so.

Best Practices

Operation Return:
A Learning Experience
1978

UNFORTUNATELY, MOST PEOPLE only become aware of fire when tragedy strikes. Operation Return, a project of the District of Columbia Fire Department, is intended to turn morbid curiosity into a vicarious learning experience. As cruel as it might seem, it is important for the fire service to capitalize on the public's "moment of interest" when a fire strikes. While the fire tragedy grabs the public's attention, Operation Return gives them that may prevent them from experiencing a similar tragedy.

In many instances, information reported about a fire sounds or reads like this: "One person died and two were injured in an early morning fire. It took twenty-five firefighters an hour to control the blaze. The cause is under investigation." The report is accurate but citizens have not received any information or knowledge that will keep them from experiencing a similar tragedy. Operation Return hopes to change that.

After being informed of a tragic fire, the department's Community Relations Fire Safety Education Unit starts collecting information about the fire. The fire investigator's report is reviewed, interviews are conducted with the first arriving company, citizens present at the time of the fire are questioned, and the fire scene is visited. After the data are carefully analyzed, a conclusion is made regarding what "specific action" might have changed the outcome of the fire. The specific action might not necessarily have prevented the fire, but it could have averted the tragedy.

Examples:

■ A smoke detector would have alerted the victim in time to escape.

■ If the security bars on the window could have been opened from the inside the victim would have escaped.

■ The occupants tried to fight the fire before calling the fire department.

■ If the victim had hung from the window ledge the ladder would have reached him in time.

■ A sprinkler system would have prevented the night club from being destroyed by the fire.

When the decision is made concerning which one of two specific behavioral objectives are to be stressed, the editors and news directors of the city's newspapers and radio and television stations are contacted. They are told the fire department will return to the fire scene at ten a.m. to explain why the fire tragedy occurred and how it might have been prevented. It is important to schedule meetings with the news media early in the morning in order to have the message aired on the evening news.

The local media have been extremely cooperative because they see the value in such reporting. It is estimated that 589,000 adults view the evening news in the Washington, D.C., metropolitan area. So far, we have documented proof of one family's surviving a fire as a result of the knowledge it gained from viewing Operation Return on a news broadcast.

While the news media are being briefed, flyers are distributed in the immediate neighborhood of the fire informing the citizens where and when the fire department will hold a meeting to explain how the fire tragedy might have been averted. The department tries to hold the meeting at a local church, hall, or neighbor's home within twenty-four hours of the fire. The meeting usually is scheduled in the evening around seven p.m. so attendance can be maximized.

One or two members of the Community Relations Fire Safety Education staff conduct the meeting. It consists of a thirty-minute lecture that explains in detail what the audience can do to prevent a similar experience. Another thirty minutes is alloted for questions and answers on any fire-related topic.

This portion of Operation Run has been well received and effective. More than five hundred people have been educated on this face-to-face basis.

The Fire Safety Education staff has averaged about ten working hours for each Operation Return conducted. This is a small investment when compared to the number of people being educated. Public fire-safety education is the key to solving the fire problem, and the best time to educate is when you have the public's attention.

Typical Operation Return scenario

8:15 am.   Community Relation Fire Safety Education Unit notified that a fatal fire occurred the previous evening. Basic information gathered: location of fire, victim's name, age, sex.

8:30 a.m. Copy of fire investigators report obtained. If report indicates natural causes we proceed; if arson or suspicious, we discontinue Operation Return because it might interfere with the case. In this case, natural causes are indicated.

8:45 a.m. News media are called and advised that the fire

department will return to the fire scene at ten a.m. to explain the fire and what might have averted the tragedy.

9:00 a.m. The Fire Safety Education Team arrives at the fire scene. They find someone in charge of the property and explain what they would like to do and ask permission.

9:15 a.m. One team member examines fire scene, the other conducts interviews. They gather information to answer the following questions.

Question	Answer
Location of victim	Bedroom floor
Fire origin	Living room
Cause	Smoking
Fire spread	Living room and hall
Who discovered fire	Person in apartment above
Action taken	Person called maintenance man to report odor of smoke. Maintenance man went to caller's apartment and determined odor was coming from apartment below. He went to the fire apartment, opened the door with a pass key and tried to fight the fire with extinguishers. The fire made him retreat. He blocked open the apartment door with an extinguisher, "so the firemen could find the fire." He pulled the interior fire alarm. He did not know the alarm was not connected to the fire department.
Who called	When the resident manager heard the alarm, she called the fire department.

Specific action that could have changed the outcome of the fire.
1. A smoke detector would have alerted the victim in time to escape.

	2. Anyone who thinks there is a fire should call the fire department.
	3. Do not fight a fire—call the fire department immediately.
	4. Close doors in the event of a fire.
10:00 a.m.	News reporters start arriving. Each reporter is given the opportunity to interview a member of the fire department and film the fire scene. By 11:30 a.m., four television reporters filmed and conducted interviews. All four television stations air the Operation Return message for this fire on 5:30, 6:30 and 11:00 news. This reaches 1.5 million adults.
Noon	Flyers posted in elevators and on each floor announcing the fire department meeting in the lobby, the next night at 7:00 p.m.
7:00 p.m.	Two Community Relations Fire Safety Education members conduct a one-hour program on smoke detectors and escape planning. Ninety tenants attend.

Ethnographic Interview of a Burn Patient
1983

TWELVE YEARS AGO I walked into a fire station and my fire-service career began. Two weeks ago I met David Anderson and after one hour of listening, I learned more about pain, the will to survive, and compassion than ever before.

The connection between the fire service and David is one that is repeated hundreds of thousands of times every year. David is a fire victim. What is it like to be a fire victim? Can we come to understand David's world?

To answer these questions and to understand the culture of the fire victim, an ethnographic interview was conducted with David. Arrangements were made with City Hospital to talk to a burn patient. David was suggested as a possible interviewee. So when his condition stabilized, and David agreed, an hour interview was conducted.

When you walk into City Hospital your first impression is that it is old and run-down, not a typical TV version of a hospital. The burn unit is on the top floor; a shaky elevator takes you to the sixth floor. When the elevator doors open, you're in the visitors waiting room with its vinyl furniture, tile floor, and cold atmosphere. As you turn right the corridor seems to narrow; there are nurses, orderlies and patients all moving around. The first nurses station is crowded with busy people. When you pass through the swinging double doors you can immediately tell you're in the patient area. All the room doors are open. It's difficult not to invade privacy by staring into the rooms.

Room 635 has two patients. You can tell one is much more severely burned than the other. White bandages cover most of his body, both legs, and his arms and head. He has a tube in his nose. The room is cramped. There is only enough room for one person to stand on either side of the bed, and you can only pass the foot of the bed one at a time.

Despite the old, run-down, cramped, disheveled impression, this is the best center in the state and a leader in the country.

Human beings must have certain things to survive: air, water, food, shelter, and a will to live. These basic human needs are present in all cultures. The needs of the burn patient are not different than the needs of any other culture. The difference lies in the acuteness of their needs.

Throughout the interview with David, three subjects were at the root of all of the experiences he talked about: pain, will to survive, and compassion. All people experience these, but being burned heightens the experience beyond our comprehension.

Life does not exist without pain. No one escapes experiencing pain to some extent. Pain varies in intensity from that which produces mild discomfort to that of intolerable agony.

When you cut your finger it hurts; you experience pain. You didn't intend to cause the pain; it was an accident. Now you must have medical treatment for that cut. A little iodine will do the trick. You know it will hurt when applied but it must be done. You experience pain again for a moment when the iodine is applied. No one experiences more pain than the burn patient. Burn patients develop a long list of words to describe the different degrees of pain they experience.

Every morning David is wheeled to the washroom for his bath. A winch is used to lift David from the gurney and lower him into the stainless steel tub filled with chlorinated water. Being immersed in the water burns terribly. David explains, "The burns get little red things on your skin. When it starts to grow back little red buds, your nerves are starting to grow back. When they scrub you down, as soon as they touch you with that rag it feels like somebody with sandpaper is ripping the meat off of you. Just to touch it is terrible." Once the dead skin is loosened it must be removed. A nurse is on either side of David. They use tweezers to pull the skin away from the body and then cut it off. If this procedure were not conducted repeatedly, infection would set in and David would die.

David uses six different words or phrases to describe different degrees of pain.

Greatest: Terrible / rough experience / really burnt bad / pretty painful / uncomfortable

Least: Feels good

The will to survive exists in all people. When that will is given up, death can not be far behind. What causes a person to have the will to live? Each person probably has a different answer. For the burn patient the need to fight to survive is paramount.

David was brought to the Burn unit with second- and third-degree burns over forty percent of his body. The doctors told his parents that he would not live until morning. They were told they could see David, but he would not be able to speak to them because he was unconscious. When his parents started talking to David he amazingly answered them. He was also alive the next morning.

The last thing David remembers after being pulled from

his burning truck is thinking, "If I die who will take care of my little boy?" That thought is David's will to survive. "Thinking about my boy kept me alive."

After twenty-three days in the hospital, four operations, and two months of daily pain, the will to survive must be constantly maintained. "Sometimes I get so depressed. But when I get like that I just call my boy and talk to him. It helps just to hear his voice."

Having a future is a major part of the survival instinct. For the burn patient, things we may consider simple become the most important. If you ask David what he is going to do when he gets out of the hospital he says, "I'm going to hug my kid, kiss my girlfriend, see my family and friends. I'm going to go home."

Compassion is defined as sympathetic consideration for others. It is a human trait and exists in all cultures. We may not always know when we are being compassionate but when we need it and don't receive it, the lack of compassion is very obvious.

Anyone who is injured in our society is given compassion almost automatically. It's probably an outgrowth of the fact that we have all received compassion when we were injured. Only those who have been severely burned can truly under stand the burn patient's need for compassion. Those who work with burn patients also have an understanding of that need. But the ability to give compassion is a very individual characteristic. David classifies medical staff by their level of compassion. "Some of them are better than others, more gentle with you, then you get the ones that are in a hurry. They don't much care how much they hurt you."

Every day David's bandages and bedsheets are changed. As he lies in bed the burns on his back stick to the bed clothing. He must be removed to replace the sheets. David can pull himself up from the bed very slowly, using the trapeze hanging over his bed. To get himself loose, it takes about ten minutes.

"Some will let you pick yourself up; others just pull you up. I've had some pretty rough ones. In fact, I had one that I asked to leave the room and not to come back. I reported her to my doctor. Every time she walks by my room now, she gives me a dirty look. That's her problem. I didn't come here to get hurt; I've been hurt enough."

The themes of pain, will to survive, and compassion are very interrelated. Suffering in some form is experienced by all. The ability to go on despite the greatest odds is constantly repeated throughout the history of mankind. The fact that people are capable of doing terrible things to one another is countered by the humanness we all share.

The world of the burn patient is not different in kind than ours but different in degree. All physical and emotional phenomena are heightened to the ultimate level.

Lessons from America's Best-Run Fire Departments

1985

WHICH FIRE DEPARTMENTS are among the top in the country? What makes them excellent? What can other fire departments learn from them?

These are the questions *Firehouse* editor John Peige asked me to research back in February in preparation for this special "In Search of Fire Service Excellence" issue.

Using the nationwide best-selling book, *In Search of Excellence: Lessons From America's Best-Run Companies* by Tom Peters and Robert Waterman Jr. as a guide, we undertook an intensive study and survey over the intervening months to find out if the eight basic management principles that account for the success of best-run companies like IBM, Hewlett-Packard, McDonald's, and 3M also account for the success of the best-run fire departments. (See "Management Principles," page 202.) The answer: most definitely *yes*. In the following article I will explain how the twenty-five excellent fire departments were identified, what measures were used to validate their first-rate status, what they are doing to maintain their position, and why the chiefs of these departments think their organizations are excellent.

Excellence based on perception

The first step was to narrow down twenty-five thousand fire departments in the country to a manageable number. This process gives us our first insight into the concept of excellence. The idea of an organization being excellent is based on the perception of those doing the judging. Peters and Waterman used an "informed group of observers" to develop their original list of companies to be studied. My group of informed observers was made up of seven people.

The group included a former commissioner of the Chicago Fire Department, deputy chief of the New York City Fire Department, the Kentucky state fire training director, the Maryland fire marshal, a training instructor, a National Fire Protection Association investigator, and a college instructor. Together they have worked with thousands of fire departments around the country and also serve as full-time or adjunct faculty for the National Fire Academy. The members were chosen because of their different perspectives on what excellence means, based on their background, experience, and education. (See "Informed Observers," page 204.)

Each was given a list from the *NFPA Fire Almanac* of eighty-seven metro fire departments and asked to check off those they considered to be excellent, using whatever definition of excellence they had. A total of thirty-eight fire departments were selected from the list. It was decided that for a department to be included, three or more members had to select it. Fourteen departments made the first step.

To be thorough, the group would need to study other than metro fire departments. Therefore, the members were asked to add to the list any fire department they considered excellent. Eleven fire departments were named by the group.

Burton A. Clark

The metro list and the random list provided our sample list of twenty-five excellent fire departments. The study sample is very representative because it includes departments from fifteen different states; all six regions of the country; and paid, volunteer, combination, urban, suburban and rural fire departments.

25 excellent fire departments

Fire Department	Survey	Fire Chief	Interviewed
Alexandria, VA	•	J. Hicks	
Bloomington, MN	•	George Hayden	•
Charlotte, NC	•	R. Blackwelder	•
Charlotte, VT	•	David Schemerhorn	•
Charlottesville, VA	•	J. Taliaferro	•
Colorado Springs, CO	•	Richard Smith	•
Dallas, TX	•	Dodd Miller	
Fairfax, VA	•	W. Isman	
Fort Worth, TX	•	Larry McMillan	•
Lancaster, PA	•	J.D. Mumman	•
Los Angeles, CA	•	Don Manning	•
Louisville, KY	•	Larry Bonnafon	•
Memphis, TN	•	Bill Burross	•
Miami, FL	•	K. McCullough	•
Midwest City, OK	•	Tom Canfield	•
New York, NY		John O'Rourke	•
Palm Beach, FL	•	Ken Elmore	•
Phoenix, AZ	•	Alan Brunacini	•
Salt Lake City, UT	•	Peter Pederson	
San Clemente, CA		Tom Daily	•
San Diego, CA		Roger Philips	
San Jose, CA		Robert Osby	•
Seattle, WA	•	C. Harris	•
Virginia Beach, VA	•	H. Diezl	•
West Sadona, AZ	•	John Olson	•

Measures of excellence

The second step Peters and Waterman used was to devise a plan to define what is exactly meant by excellent. What do we measure to determine if a company is excellent? They used financial data to measure growth, long-term wealth, return on capital, and sales. Fire departments do not have profit-and-loss statements or price/earnings ratios, so we had to develop other measurable criteria.

This problem was presented to a group of thirty-five fire executives from around the country. The group of executives was divided into three small work groups. Each group was asked to "develop a list of six measurable criteria" they would use to determine whether a fire department was excellent. The results of the three groups were almost identical. The group identified six measurements of excellent performance over a three- to five-year period of time. To be excellent, a fire department must show a trend in six areas:

- Decrease in fire loss
- Decrease in firefighter injury
- Financial responsibility
- Increase in performance standards
- Increase in amount and type of service offered
- Decrease in response time

To collect these data, we sent out surveys to each of the twenty-five fire departments. Twenty-one were returned in time to be included in the study. Each department's data were analyzed individually to identify trends, then the departments were compared to identify top performers. This is a very rigorous evaluation method. The results show that eighty-nine percent of the departments were top performers in three or more of the categories.

Top performers

Departments	Fire Loss Decrease	Firefighter Injuries Decrease	Performance Standards Increase	Services Offered Increase	Cost Per Capita	% Jurisdiction Budget
Alexandria, VA		●	●		●	
Bloomington, MN	●	●			●	
Charlotte, NC		●		●	●	
Charlotte, VT		●		●	●	
Charlottesville, VA	●	●	●			●
Colorado Springs, CO		●	●	●		●
Dallas, TX		●		●		●
Fairfax, VA	●	●	●		●	
Fort Worth, TX		●	●	●	●	
Lancaster, PA		●			●	●
Los Angeles City, CA	●	●	●	●	●	
Louisville, KY			●	●	●	●
Memphis, TN	●	●			●	
Miami, FL	●	●	●	●	●	
Midwest City, OK		●	●	●		
Palm Beach, FL	●	●	●	●		●
Phoenix, AZ	●	●	●	●		
Salt Lake City, UT	●	●	●		●	
Seattle, WA	●	●	●	●	●	
Virginia Beach, VA	●	●				
West Sadona, AZ	●	●				

Survey methodology

Fire loss. There are four measures of fire loss: death, injury, dollar loss, and the number of incidents divided by the dollar loss to represent dollar loss per alarm. These data were collected over a five-year period from each department.

Sample fire loss data

Year	1980	1981	1982	1983	1984
Death	14	6 ▼	5 ▼	4 ▼	2 ▼
Injury	32	30 ▼	29 ▼	10 ▼	36
$ Loss	1.080M	991K ▼	761K ▼	603K ▼	1.730K*
$Loss/Alarm	108	99	76	67	173

Dollar loss was adjusted for inflation using Consumer Price Index

The arrows indicate decreases from the previous year's number. If a fire department were perfect it would have sixteen decreases. The actual number of decreases can be divided by sixteen to give the department's fire loss "batting average," as in 13/16=.810. It was determined that for a fire department to be considered a top performer in the fire-loss category, it needed at least a .500 average.

Firefighter injury and death. The same averaging method was used to determine a fire department's trend in decreasing injuries. It is interesting to note that all fire departments surveyed had at least a .500 average.

Financial responsibility. Two data items were used to identify financial responsibility: cost per capita (CPC) to show cost effectiveness, and percentage of fire department share of overall jurisdiction budget, which demonstrates financial support for the fire departments. The CPC was adjusted for inflation using the Consumer Price Index for purchasing power. The first-year numbers were compared to the last-year numbers, and the percentage of change, plus or minus, was noted. For example, the CPC in 1980 was $17.88, which equals $7.26 when adjusted for inflation. CPC for 1984 was $24.83; the adjusted CPC is $8. This equals a ten percent increase in the CPC over the five-year period. To be included in the top half, a department had to have a CPC increase of ten percent or less over the evaluation period.

The number of times a fire department's percentage of the total jurisdiction budget increased was also measured. To be considered a top performer, the department had to receive a percentage increase at least half the time.

Performance standards. The survey asks the department to give a narrative description of how the performance standards of its personnel have changed over the past five years. The number of changes listed by each department was counted. There was no attempt to weigh one change over another. As with the other measures, to be a top performer in this category the department had to be in the top half, which meant nine or more performance standards changed over the five-year period. Grouping the changes gives us a clear picture of what the excellent fire departments are doing.

The overwhelming change is physical fitness, with the majority of the departments having mandatory physical fitness programs, including the Bloomington, Minnesota, Volunteer Fire Department. Some departments have been involved in fitness programs for a number of years. The Salt Lake Fire Department has been involved in long-term research with the LDS Fitness Institute and is currently developing a complete EKG treadmill test for firefighters. Departments are also beginning to be concerned about mental health issues. The Fort Worth Fire Department has a team of stress-intervention specialists to help firefighters and their families cope with job-related stress.

NFPA professional qualification standards were the next most often mentioned performance-standard change. Firefighter I, II, and III were identified as being used to certify personnel. About half the department also identified some level of officer training, certification, and development program.

Very special programs were developed for firefighters in Texas, Virginia, and California. The Dallas Fire Department has a management rotation program, in which all chief officers are rotated through other city departments. The Fairfax County Fire Department has a foreign exchange program, which allows firefighters to trade places with their foreign counterparts. A department-sponsored bachelor's degree program was established by the Los Angeles City Fire Department in conjunction with the University of Redlands.

The fire-prevention divisions of the excellent fire departments are also getting a lot of attention. Personnel are being qualified under various local and state systems as prevention specialists and code enforcement officials. Fire-prevention performance standards are being included at the company level because they are conducting more and more prevention activities.

Finally, it is interesting to note that almost all the departments cited participation in National Fire Academy courses as a part of their professional development programs.

Services offered. These were measured by counting the number of changes identified by the department. Again, there was no attempt to weigh the changes. To be a top performer, the department had to list ten or more changes in this category.

What services are the excellent fire departments delivering? Most often, fire-prevention activities, followed by emergency medical services and supervision.

Fire prevention activities accounted for the most change. The types of change included starting company-level inspection programs and hiring fire protection engineers. Some of the unique changes include the following:

- The Miami Fire Department took over the city building and zoning department, which then created the Department of Fire, Rescue and Inspection Services.

- The Seattle Fire Department instituted a program of checking all businesses from current business licenses during annual fire inspections. This service netted the city $65,000 in uncollected business taxes the first year. Residential smoke-detector programs were conducted by most of the departments.

- The Fort Worth, Texas, Fire Department participated in the installation of 9,500 smoke detectors—just as impressive as the twenty-four-member Charlotte, Vermont, Volunteer Fire Department visiting every residence in the community to be sure each had at least one smoke detector. Major increases were also made in public fire and burn education programs. The programs identified most often were Learn Not To Burn, Juvenile Fire Setters, and Home Inspections.

EMS was the second biggest change area. These changes covered a broad range, which included starting a first-responder program with an engine company, and testing new advanced life support equipment, as in the sophisticated emergency medical systems of Miami and Seattle.

The changes in EMS delivery seemed very diversified and indicated an effort to deliver the desired level of service the individual community wants. For example, the Phoenix Fire Department has five paramedic rescue units and thirteen paramedic engine companies. The Palm Beach Fire Department carries medical data cards for residents on its two medic units.

Finally, the overriding change in the suppression field was the creation of hazardous material response units over the past five years. Though it is last, hazardous materials are not taken lightly. According to Chief Manning, Los Angeles City: "Hazardous materials are the single greatest threat to the community."

Response time. Unfortunately, not enough of the departments surveyed could supply the data needed to show changes in responses from time of ignition, so this category was dropped from the analysis.

Why are they excellent?

To answer this question, we conducted interviews with the fire department chiefs, asking them one question: "Why do you think your fire department is excellent?" Their responses were categorized under each of the eight principles. Five of the eight principles were given by all of the chiefs as the reason for their organization's being excellent. They also identified a new principle.

Bias for action. Excellent fire departments are willing to try new things and new methods. They are change-oriented and take risks. They are also supportive when things fail. Noted Chief McMillan of Fort Worth: "Fire departments need to change as the times change and the needs change." Chief John O'Rourke of New York City echoed this action-oriented approach when he said, "We don't stand still. We don't sit back on our laurels ... we try to move ahead by trying new things or taking a pragmatic approach."

The chief must have a personal commitment to this action orientation. Said Chief Tom Daily, San Clemente, "I think I have to keep an open mind on anything that is suggested by anybody . . . don't be afraid to try something new."

For this action bias to work, there must be support for it at all levels. Chief Roger Philips, San Diego, explained this support when he said, "There needs to be a willingness and a freedom to try new things and then, if they don't work, to not punish the person." Chief J. Taliaferro of Charlottesville explained the support he receives from his city manager as follows, "He encourages you to be innovative, to take some risks, to try something different; at the same time, he supports you if things don't go right."

Listening to the public. Fire departments belong to the public, so the public has a right and responsibility to be involved. Excellent fire departments seek involvement from the public and as a result have strong community support.

It is obvious why a business must keep in touch with what its customers need. But why would a fire department listen to the public? Chief John Olson, West Sedona, Arizona, gives the perfect reason. Says he: "When we put our organization chart together, above the elected officials is the electorate—they are in our chain of command." The responsibility to the public is summed up by Chief Roger Philips, San Diego, California: "That is the citizens' money we are spending. We must be concerned with balancing the budget and being cost-effective."

Close to the customer means being involved all the time, not just when the alarm bell goes off. Chief Tom Canfield, Midwest City, stays in touch with all that is going on in the city. "In my six years as chief I have attended ninety-eight percent of the council meetings . . . ninety-five percent of the time there is no fire department business going on at all, but every time the council looks up it sees me—not just when I need something."

It doesn't matter what size the fire department is—it must listen to the public. In New York City, the department listens to its local neighborhoods through community boards. Chief John O'Rourke said, "When they [the neighborhood] tell us something that they want or need, we evaluate it and make every effort to give it to them."

Chief Bill Burross, Memphis, gave a clear example of how local involvement is a two-way street. He explained, "The whole neighborhood became involved in a fire station beautification project. They planted grass and flowers and shrubs. They feel it is their fire station; they want to be part of it and involved with it."

People: The most valuable asset. A fire department is only as good as its people. Excellent departments believe this and do everything possible to develop and utilize their members' talents throughout the organization.

"Our people are not the enemy," said Phoenix Chief Alan Brunacini. "Our rival in the system is service delivery. It isn't the people, it isn't the union, it isn't the city, it isn't the citizen. It's the fact that there are buildings on fire and there are people who are having various kinds of emergencies. That is what the challenge really ought to be."

Excellent fire departments truly value their people. In San Diego, according to Chief Philips, "What we try to instill in everyone—secretary, mechanic, firefighter, and deputy chief—is that his job is to serve the public, that all jobs are important to the overall effort of the fire department. It's appreciated and it's valued."

The chiefs expressed a sincere level of concern for their people. Chief Taliaferro of Charlottesville put it simply: "We try to put a lot of emphasis on the little things with our people—dealing with the little complaints that people have. It's the small things that make the difference." Chief Ken Elmore, Palm Beach, expressed his deep commitment to developing his people: "The way I get my kicks is seeing the department grow and improve. I'm not talking about bricks and mortar. I'm talking about people. That's what the department is. If they can grow and improve then that gives me my kicks and I have fun at it."

The excellent fire departments are committed to using their most valuable assets. Unfortunately, it is not always the case, according to Chief Larry McMillan, Fort Worth: "My concern throughout my fire service career has been the underutilization of its talent . . . my goal is to utilize that talent, put it to work, and provide the service we are capable of giving."

Autonomy. Excellent fire departments realize that the person who knows how to do the job best and how it could be done better is most likely the person doing the job. Management must foster individual creativity and innovation. This translates into letting people do their jobs, delegating and realizing that the chief cannot do it alone.

"It's not a one-man show," said Seattle Chief Harris. "It's not really up to me to come up with all the creative and innovative ideas, but it is up to me to seek input from the rank and file . . . on how to do a better job." Chief Brunacini, Phoenix, says, "I'm the easiest person in the world to help."

The Charlotte, North Carolina, Fire Department puts the management responsibility where it belongs—with the company officer "that allows them to be the boss, to be the manager, to be the supervisor, to be the chief," says Chief Blackwelder.

Chief Elmore, Palm Beach, gives us insight into why we need to encourage autonomy. "They want to see their ideas work. That is what makes it work."

Values. The leadership of excellent fire departments, especially the chief, plays a significant role in shaping the value system of the department. For better or for worse, the chief sets the tone for the attitude of the organization.

When asked what values they try to instill in their

organizations, the chiefs gave predictable answers: honesty, loyalty, high expectations, pride, respect. But those were not the gut-level feelings that were evident in every chief. It was the joy, the enthusiasm, the excitement in their voices when they talked about their departments during our telephone interviews. That is what they instill in their organizations. From the tone in their voices, they seemed to be a group of rookies on graduation day. These chiefs were at the beginning of their careers, not the end.

Training. Excellent fire departments encourage individual, group, and organizational growth. They believe they can be better and know they must be better to meet future challenges.

The chiefs gave a lot of the credit for excellence to their training efforts in terms of quality, quantity, and commitment. From basic recruit training to officer development, training was equated with excellence. Chief Elmore, Palm Beach, said, "We are not bricklayers. We don't practice all our needed skills every day. We must constantly drill and train to stay competent." Noted Chief Don Manning of Los Angeles City: "I'm preparing the next chief to do a better job than me."

What wasn't mentioned. A very interesting fact is that none of the chiefs equated their fire department's firefighting capabilities with its excellence. Firefighting was never mentioned. The chiefs credited their excellence with their prevention program, EMS delivery, public education, and community involvement programs. Chief Brunacini came the closest to mentioning firefighting when he said, "When the bell hits and it's showtime, the people perform." Chief Manning did talk about the LAFD's Incident Command System, but said that "the professional and personal care we give to the victims is what makes us outstanding." It may be that in the excellent fire departments, firefighting capabilities are a given. Excellent performance has become the norm.

Three of the principles identified by Peters and Waterman

were not identified. First, excellent fire departments are not sticking to the knitting; they are becoming more and more diversified in an effort to meet the needs of the public. Two principles dealing with form and loose/tight controls could not be identified using the interview technique; on-site observation would be necessary.

Conclusion

This, we feel, is an interesting and viable way to measure changes occurring in fire departments, not just for the twenty-five departments mentioned, but for all of us in the fire service. There is no magic to being an excellent fire department, because all of us can use these basic principles. A fire department's success is limited only by its personnel. And we all know that fire service personnel can do anything, if they want to.

Management principles

"Our findings were a pleasant surprise. The project showed, more clearly it than could have been hoped for, that excellent companies were, above all, brilliant in the basics. Tools didn't substitute for thinking. Intellect didn't overpower wisdom. Analysis didn't impede action. Rather, these companies worked hard to keep things simple in a complex world. They persisted. They insisted on top quality. They fawned on their customers. They listened to their employees and treated them like adults. They allowed their innovative product and service 'champions' long tethers. They allowed some chaos in return for quick action and regular experimentation." That's the conclusion Thomas Peters and Robert Waterman Jr. came to in their best-selling book *In Search of Excellence, Lessons From America's Best-Run Companies*. You may notice a similarity in their book title and the title of this article. The similarity is not an accident, because the purpose of the research on which this article is based was to find out if the same basic management principles that account for the success of the best-run companies also account for the success of the best-run fire departments.

The Peters and Waterman book, which served as the basis for this article, concluded that these basic management principles account for the success of the best-run companies:

One: A bias for action—a preference for doing something—anything—rather than sending a question through cycles and cycles of analyses and committee reports.

Two: Stay close to the customer—learning preferences and catering to them.

Three: Autonomy and entrepreneurship—breaking the corporation into small companies and encouraging them to think independently.

Four: Productivity through people—creating in all employees the awareness that their best efforts are essential and that they will share in the rewards of the company's success.

Five: Hands-on, value-driven—insisting that executives keep in touch with the firm's essential business.

Six: Stick to the knitting—remaining with the business the company knows best.

Seven: Simple form, lean staff—few administrative layers, few people at upper levels.

Eight: Simultaneous loose/tight properties—fostering a climate where there is a dedication to the central values of the company combined with tolerance for all employees who accept those values.

"We [Peters and Waterman] should note that not all eight attributes were present or conspicuous to the same degree in all of the excellent companies we studied. But in every case at least a preponderance of the eight was clearly visible, quite distinctive. We believe, moreover, that the eight are conspicuously absent in most large companies today. Or if they are not absent, they are so well disguised you'd hardly notice them, let alone pick them out as distinguishing traits. Far too many managers have lost sight of the basics, in our opinion: quick action, service to customers, practical innovation, and the fact that you can't get any of those without virtually everyone's commitment."

Informed observers

The following are the "informed observers" who worked on the "In Search of Excellence" project:

William R. Blair: Thirty years in the fire service, former commissioner of the Chicago Fire Department, retired assistant chief of Los Angeles City Fire Department. Presently in charge of the National Fire Academy Emergency Incident Policy and Analysis program. A.A.S. in Fire Service.

Charles J. Burkell: Eight years in the fire service as firefighter, training instructor, EMS manager and safety consultant. Presently in charge of the National Fire Academy Education program. B.A. in Fire Technology, B.S. in Technical Education, M.B.A. candidate.

Steven W. Hill: Fourteen years in the fire service, former National Fire Protection Association investigator and Washington State assistant fire marshal. Presently in charge of the National Fire Academy Management Technology program. B.A. in Fire Protection.

Bruce W. Hisley: Twenty-four years in the fire service, retired fire marshal of Anne Arundel County, Maryland, Fire Department. Presently on the NFA faculty in the Community Research Preservation Program. Member of NFPA Standards Making Committee. A.A. in Fire Service.

Lucien P. Imundi: Thirty-five years in the fire service, retired deputy chief of New York City Fire Department. Former faculty member of the National Fire Academy Incident Command program. Presently founder and director of the National Fire Service Command Logistics Institute. B.S. in Behavioral Science.

Jan D. Kuczma: Five years in the fire service, former chemistry teacher. Presently on the faculty of the NFA in the Hazardous Material program. B.A. in Chemistry/Biology.

A. Don Manno: Eighteen years in the fire service, former director of Fire Training, Kentucky. Presently in charge of the NFA Community Resource Preservation program and member of NFPA, MAC Committee. B.A. in Psychology.

Tests Are An Important Fire-Rescue Service Tool
1995

ACCORDING TO THE American Psychological Association, "Educational and psychological testing represents one of the most important contributions of behavioral science to our society . . . the proper use of well-constructed and validated tests provides a better basis for making some important decisions about individuals and programs than would otherwise be available."[1]

Three important personnel questions that face the fire-rescue service are:

1.) Did we hire the right person?
2.) Did the person successfully pass required training courses?
3.) Did we promote the right person?

The word *right* is a very subjective term, but tests are supposed to be objective. For many years, the career fire service has used testing to help justify the decision to hire and promote individuals. Both the career and volunteer service have used testing to determine if firefighters have successfully mastered the knowledge, skills, and abilities necessary to perform the job. What validity and reliability procedures are used by the fire service to ensure the effectiveness of their examinations?

Testing will continue to play an important role in personnel selection, promotion, and certification. As the number of entry-level positions are decreasing, the pool of applicants is increasing. Competition for promotion is also increasing as the job is becoming more complex. The American Psychological Association states, "Competent test use can make significant and demonstrable contributions to productivity and to fair treatment of individuals in employment settings. Among available alternatives, tests are the most valid and the least discriminatory personnel decision aids available."[2] Tests are tools that the fire service can use to help reduce the subjectivity, biases, and opinionated aspects of the decision-making process.

The quality of decisions based on tests are only as good as the qualify of the test in terms of the instrument's validity and reliability. Validity questions ask, "What does the test measure?" and reliability questions ask, "Is the test consistent?"

"Reliability is concerned with the stability of test scores—does not go beyond the test itself. Validity, on the other hand, implies evaluation in terms of outside independent criteria."[3]

Finally, a test, like a ruler, a scale, or an EKG, measures phenomena. The ability to measure human understanding of the physical and psychological world is directly correlated to the validity and reliability of the instruments used. There is a wealth of knowledge on test development. The fire service relies heavily on testing as part of its decision-making process. Therefore, we need to apply the available knowledge on testing to our testing instruments so we can answer the questions:

1.) Are fire service tests valid and reliable?

2.) Will decisions based on testing be the "right" ones? 3.) Will we pick the "right" person for the "right" job? These are important question to answer because, at some time, we will all be test developers, test takers, or test-result users. Good luck on your next test!

References

[1] American Psychological Association, *Standards for Educational and Psychological Testing* (Washington: American Psychological Association, 1985), 1.

[2] *Ibid*, p. 59

[3] Henry E. Garrett, *Statistics in Psychology and Education* (New York: David McKay Company, 1971), 360.

Test Your Legacy Potential
2002

ALL OF US have benefited from the work of the twenty fire-service legends that most shaped our discipline in the twentieth century. By examining their careers, we may be able to identify common characteristics among them that helped them make such significant advances in the fire and rescue services.

This information can be useful in our own professional development, allowing us to create organizational environments that help others contribute to their highest potential. By studying those who came before us, we honor the past, we celebrate the present because of how far we have come, and we can believe in the future because of the legacies we are creating.

Fire service legends

J. Ray Pence
Ralph Scott
Lloyd Layman
Fred Sheppard
Robert Gain
Percy Bugbee
James Meidl
Warren Isman
William Clark
Gordon Vickery
Keith Klinger
David Gratz
Howie McClennan
Warren Kimball
Lou Witzemen
John Bryan
Francis Brannigan
James Page
Alan Brunacini
Ronny Coleman

Will you leave a legacy to the twenty-first-century fire service? Take this test to measure your legacy quotient.
- Are you the fire chief of your organization?
- How many years in the fire service do you have?
- How many times have you been published?
- What is your highest academic degree?
- What subject area are you considered an expert in?
- What major change did you champion?
- How many times have you lectured to a national audience?
- Do you plan to work beyond the year 2050?

By using the careers of our twentieth-century legends as a baseline we can get an idea of what we will need to do to leave our legacies to the twenty-first-century fire service.

Are you a fire chief? Thirteen of the twenty legends were fire chiefs during their careers. There are at least five reasons why being fire chief is important to your legacy. First, a fire chief has position power; the chief is in a critical leadership position. Whether or not you use the position effectively is largely up to you. Second, as chief you have organizational mass to use. It's difficult to be an artist if you don't have material to work with. Next, your organization can be a laboratory environment or a studio in which to practice your art and science. Fourth, as chief you have the opportunity for success as an individual and, more importantly, through your organization. The challenge is that you aren't the one who determines if you're successful or not. That leads to the last reason: visibility. Being the chief puts all eyes on you. Everyone knows if you win, lose, or draw.

So if you're the fire chief give yourself ten points on the legacy scale. If you're not chief, you get five points because seven of the legends were successful without five bugles. Remember, all of us can be the chief of our own destiny.

How many years of service do you have? Our twenty legends have an average of thirty-four years of service. There's a correlation between legacy and longevity: The longer you're around, the more time you have to get it right. Also, there's a natural connection between years of service and respect, but keep in mind that respect is a two-way street. If senior people forget this they can be infected with dinosauritis, and the only vaccine is wisdom. I don't know how you acquire wisdom, but all twenty legends had lots of it.

You get ten points if you have thirty-four or more years of service and five points if you have less than thirty-four years. If you're a dinosaur you get zero points.

How many times have you been published? Astonishingly, our legends were published an average of fifty-eight times. I use the word astonishing because as a discipline we're not prone to writing and being published. This group of people led the literature of the twentieth-century fire service and helped prove the pen versus sword concept.

For example, if you do great work or learn an important lesson in a near-fatal experience but don't write about it, your new knowledge is lost to everyone but you. Writing forces you to clarify ideas because until an idea is put on paper, it has no substance. Being published communicates your ideas to large numbers of people regardless of geography or time.

Finally, writing and publishing put your ideas up for critical review. Those who read it can accept, reject, ignore, or build on your work. All of us are building on the work of these twenty individuals today.

You earn ten points if you have been published fifty-eight or more times, eight points for thirty or more, six points for twenty or more, and four points for the rest of us because we have all written something: a desk journal, a fire or EMS report, or an accident report. Writing is so important that you can't be a professional without doing it.

How much education do you have? Twelve of our legends held academic degrees, two at the doctoral level. Fire-service degrees weren't available to most of our legends.

They learned a doctrine that they applied to the fire service. An academic degree gives you intellectual scope. All of those liberal arts courses you were required to take exercised your mind and built brain muscle. College serves as a place for idea incubation: All you learn stays with you, even if only in the unconscious mind. An academic degree gives you professional credentials in the fire service and in general society.

You get two points for earning an associate's degree, five points for a bachelor's, nine points for a master's, and ten for a doctorate. In the twenty-first century, all legends will have an advanced degree.

Are you an expert? All of our legends can be classified as experts because the rest of us look to them for guidance. Expert power is influential and not dependent on age, rank, or organization size. If you're the leading or only expert in your field, people will beat a path to your door. Being an expert gives you leadership opportunity because you can demonstrate your knowledge, skills, and abilities far beyond the topic at hand. In this position you cultivate followers not because you're the boss, but because you can help.

If you can identify what you're an expert in, give yourself ten points. Now go ask three other people what they think you're an expert in. If all four of you agree, you get twenty points.

What major change have you championed? All of our legends were paradigm pioneers. They took new ideas and made them work. The rest of us adopted their new ways, and they became the norm. We all do our job differently today because of the vision these men had. I can't score this question for you because there's no set way to become a leader of fire-service change. Becoming a paradigm pioneer is the result of a passion for what you do and a belief in continual improvement. Only you can score your passion and belief. All twenty legends had a lot of both throughout their careers.

How many national lectures have you delivered? All of

our legends were world-class public speakers. From the beginning of time, those who capture our attention when they speak give us a vision of the future, inspire us with their passion, and motivate us to action with their rhetoric. The content of the message is important, but your ability to deliver it is critical—you can't light a fire with a wet match. Public speaking is a very powerful tool, but it's also the number-one fear in adults. To get over your butterflies, take a public speaking course or join Toastmasters.

Give yourself ten points if you have lectured to national audiences twenty-one or more times, eight points for twelve or more times, five points for six or more, and two points for one or more. You get one point even if you have never spoken in front of an audience, because all of us have talked at the kitchen table and some of our greatest debates take place there. We need to move these discussions to center stage.

Do you plan to work beyond 2050? Sixteen of our legends did their most significant work after 1950. The people who work at the beginning of a new century have a more difficult time of being remembered at the end of the century. We must write our history because most of the fire-service legends of the twenty-first century are not born yet. Our work must give them a sound foundation to build on. Give yourself ten points if you plan to work beyond 2050. The rest of us get five points and have to work harder.

Now add up your score. What does it mean? Only you can determine how much you will invest in the future, and only your posterity will determine the value of your legacy.

Legacies begin in recruit school. Your job as chief is to create an environment to release the human potential. What can you do to grow legacies? Let others be in charge because leadership is found though the organization. Thinking in the long term will help others and the organization keep looking to the future even in the most difficult of times.

Encourage people to get published; having a byline is a

great incentive to write. Get a higher degree and encourage entry-level personnel to get them as well. Expect everyone to be an expert on something because we need it and it's good for the individual.

Once each person achieves expert status, look for the paradigm pioneers. These are usually individuals who keep trying to make things better. They may be the ones who drive you crazy and push your buttons because they don't take no for a answer. Learn to cherish them.

Finally, give others the opportunity to tell their story to an audience. Let a rookie give a talk at the next senior staff meeting or invite a captain to speak at the county chief meeting.

Welcome to the twenty-first century. We thank the twenty legends of the past century for getting us to where we are today. We promise to be good stewards and contribute to the legacy by investing in the future.

Am I a Competent And Courageous Firefighter?
2002

THIS IS A QUESTION I have been asking myself for thirty-three years, and I'm still challenged to be completely honest. I strive to be, but I know I do not always meet the mark. The art and science of being a firefighter is a life-and-death vocation that requires a two hundred percent proficiency level of performance. We must perform each task one hundred percent correctly, one hundred percent of the time, because your life or someone else's life depends on it. Okay, you say, but what does competency have to do with courage? You must do the right thing all the time—even when others around you are not. That takes courage.

Since the terrorist attacks of 9/11, events we all vicariously experienced, the fire service now has a new understanding of what possible dangers we face. I pray that you and I are never faced with a World Trade Center event. Those firefighters were competent and courageous that day, the ones who died and the ones who survived. My job and your job every day is to concentrate on the probable to ensure our competence and courage.

When you read the NIOSH Fire Fighter Fatality Investigation and Prevention Program reports, in every case the victims and or others at the incident were not following some part of our fire-service doctrine (standards, training, SOPs, good practice). In addition, heart attacks and vehicle accidents are still the number-one and number-two cause of firefighter fatalities. Do you have a medical examination every year? Do you wear your seatbelt going to and returning from every call? Jet fighter pilots do.

Consider all the tasks a firefighter needs to do every day. When showing up for your shift, the first thing I was taught to do was check your mask. This is a simple thing, but how many SCBAs on apparatus right now are inoperable or should be put out of service: .01%, .05%, 1%, 3%, 5%? What is the acceptable number of SCBAs being inoperable? The answer is zero, if it is your mask. For example, on one shift I was detailed during the middle of the tour for a few hours to another engine. When I got there, the firefighter I was replacing had already left. This department had day and night shifts, each twenty-four hours. Since I was four hours into the night shift, that SCBA should have been checked two times that day. My check was to be the third. You know want happened. The mask was inoperable. The harness arrangement made it impossible to put the SCBA on. This was a slow company, so they had not just gotten back from a fire. I was vocally upset and mumbling as I fixed the SCBA. The officer overheard my comments. He agreed with my discontent. The drill that night was on inspecting SCBAs. The engine and truck crews did not have any kind words for me.

Here is another example. I was teaching an instructor trainer course. One student decided to use SCBA inspection as his practice lesson. He bought in four SCBAs from his department's apparatus. He presented all the inspection procedures and then demonstrated. To his surprise and embarrassment, two of the four bottles were out of date for their hydrostatic test. Did you ever find a SCBA inoperable? More important, did you ever leave an SCBA inoperable?

I know how some of you are reacting to this. "Clark, you are nitpicking; every little thing does not make a big difference." You are right; one little thing may not make a big difference, but a lot of little things can add up to an error. This can lead to a mistake that results in an accident that becomes a tragedy. The problem with the "little-thing" argument is that we, you and I, do not know in advance what little thing in the chain of events will turn the mistake into an injury or fatality.

Take seatbelts as an example. I know this will come as a shock to all the fire chiefs reading this, but your fire department does not have one hundred percent compliance, one hundred percent of the time with the policy that requires seatbelts to be used whenever the apparatus is in motion.

When I watch NBC's *Firehouse* or Hearst Entertainment's *The Bravest*, I look to see how many firefighters are using their seatbelts. Not many. I told one firefighter I was riding with, "Put your seatbelt on." His reply was, "I choose not to wear my seatbelt." My reply was, "It is not your choice. I will file charges if you do not comply." He did. I ride out with fire departments all over the country. I estimate that seventy-five percent of the time I am the only one wearing a seatbelt. Only once did an officer reinforce that he expected me to put my seatbelt on. I was riding with a battalion chief in Indianapolis, Indiana. The chief's aide, Lieutenant Sonny Rideout, explained my "seatbelt-wearing duty" to me. I thank him. I told him I would tell his story. Thanks, Lieutenant; it takes courage to make firefighters wear their seatbelts.

In another situation, I was visiting the Lehi, Utah, fire department. It is a call department, so only one firefighter was there. We were talking when the tones went off for a car fire on the interstate. I asked if I could ride along; the officer said okay. I walked past the driver, we looked at each other, and I climbed into the cab. The driver/engineer, Captain Carry Evans, turned and announced to everyone, "Clark is here—everybody put your seatbelt on or he will talk about us." Everyone in the firetruck was stunned, mostly me; we all complied. It turned out that Captain Evans recognized me from a lecture where I talked about seatbelt usage. Thanks, Captain. Competence and courage are a powerful combination.

Competence and courage do not get any easier as you get more trumpets. The lack of them just becomes obvious to more people. As a new chief officer, I used the same car all the time. One shift I got a different car I had not used before. I checked my mask and looked at the radios; they looked different. But I was sure I could figure them out, so I did not ask for any instruction. I did not have the courage to admit I was not one hundred percent competent in using the radios. Ten minutes later, I was responding on a building fire. I was checking on the air, but did not get any reply. Communications called for me several times, and I kept replying. I thought the problem was due to poor repeaters and radio traffic, as often occurs. Then the nice dispatcher from another county explains to my communications, and to everyone within a hundred miles, that I was on the wrong radio. When I got to the scene I used the portable. I was one hundred percent sure it was the right radio. Now that everyone knew that I was not one hundred percent competent in using the radio, I asked for and received instructions.

Competence and courage does not stop when you leave the firehouse. I visited my daughter and three grandchildren at their new apartment. The kids were home, ages sixteen, nine and five. Gunner, the five-year-old, immediately took me to his bedroom. He wanted to show me the stars and clouds his mom painted on the ceiling and how they glowed in the dark. Great stuff. Looking at the ceiling, I noted the smoke detector outlet plat, but no smoke detector. The other bedroom was also missing a smoke detector. There was one on the ceiling between the kitchen and living room. Upon visual inspection, there was no light coming from the "on" indicator. I removed the smoke detector and found that it was not connected to AC power and there wasn't a battery. My daughter was home by now; I repressed my concern. My mind flashed back to January 2001 and a Delaware house fire where eleven members of one family died. Three generations gone because the smoke detectors were not working. I got Strider, the nine-year-old, to take a nine-volt battery from his Hot Car remote control; we put it in the smoke detector. My daughter is an electrician; I had her plug the detector into the AC. She did not know it was not working. I have not been very competent in convincing my daughter of the importance of having working smoke detectors. I called the fire department to let them know that the smoke detectors in that apartment complex were installed wrong. They are too close to the kitchen, so people disconnect them. I am sure they false-alarm all the time.

Recently, I was at a high-level non-fire-service government meeting as an observer. Near the end of the meeting, an announcement was made. "The fire alarm system will be worked on today, so when the alarm goes off, ignore it and do not leave the building." The hairs on the back of my neck stood up and my heart began to race, but I did not say anything. I rationalized I was just a visitor—the truth, no courage. I could not let it go. After the meeting I explained my concern to the meeting leader. He immediately understood and realized the potential danger his directions had created. He went to the fire-alarm contractor and discovered that the system would acutely be out of service. This was not known by any of the building safety personnel, so the building was placed on fire watch. Opportunities to be competent and courageous abound.

Under life-threatening situations, firefighters are expected to be at the journeyman level of performance at many tasks. Most of these skills we do not practice every day, even at the busiest fire stations in the country. As professionals, we have a duty even when we are not at the station to do what we know is right. I strive to do my best all the time. I am grateful to be part of a team that helps me be competent and courageous when I fall short. It is my responsibility to determine how I can keep improving, so I can do better the next time. This requires life-long education, training, practice, and research. Fire-service doctrine is designed to help us to be competent and courageous firefighters. Competence is a matter of doing the right thing, even the littlest thing, all the time. Courage is a matter of doing it right, even when you are the only one who does.

I keep working at being a competent and courageous firefighter. You are welcome to join me in this quest.

Train the Way You Fight—Fight the Way You Train
2003

I HAVE HEARD THE mantra "Train the way you fight—fight the way you train" from day one in the fire service. I have even said it from time to time. It just dawned on me that this could be one of the reasons firefighters get injured and killed at fires.

Recently one of our firefighters was telling me about a training experience at a house-burning drill. Craig explained that the instructor had the crew crawl down a hallway while fire was rolling over their heads at the ceiling level. When the nozzle person asked if they should hit the fire the instructor's response was, "No, crawl down the hall and get into the room on fire." When the attack crew was in the fire room the instructor ordered the nozzle to be opened, but only long enough to knock the fire down. "Don't drown it because we have to set it on fire again." I asked Craig if the room was vented. He said no, the windows were boarded up and were only opened a little at the top.

We have all been a student in a training exercise similar to this during our tenure in the fire service. Some of us have even taught it just this way. However, the question is this: Is this how we want our firefighters to practice in the real world? The answer is *no*! Our justification for teaching this way is to instill confidence in firefighters, teach them to feel the heat, instill aggressiveness, learn to get close to the fire, be macho, and get your helmet burned. Sounds like good logic, but what we are teaching is in violation of some basic firefighting rules. Do not go past any fire without putting it out. Do not crawl under fire. Flow enough water to cool the box and put the fire out. Ventilate early and often to let the heat, flammable gas, and smoke out.

So why are we so surprised when firefighters get injured or killed in unvented buildings with no water flowing? Maybe we taught them to do that.

Now when these firefighters get to be officers the "get-in-there-and-get-it" mentality does not go away. At a recent multiple-alarm fire in a large food store I had command of Division 3, the rear of the building. We were operating in a defensive mode after the steel trusses failed in an apparent metal deck roof fire. Prior to the collapse of the roof, the evacuation had been ordered early on in the fire. We had large lines in the back doors of the building and trucks were being positioned for ladder pipe operations. A chief with more trumpets than me arrived on the scene and I briefed him on the current fire conditions. Despite the fact that he had the ability to assume the division command at this incident, he allowed me to continue. I next noticed him with a crew at the back door, and they were putting their facemasks on. So I approached them and asked what they were doing.

The chief told me he was sending them into the building twenty to thirty feet so they can hit the fire better. The crew was eager to get in there so they could be real firefighters.

I said, "Chief, I don't think that is a good idea."

He replied, "Don't worry; they are only going in twenty to thirty feet.

I said, "Chief, the evacuation order has already been given."

"Don't worry; it's okay," he said. My final statement was, "Chief, you have to tell command over the radio that you are sending firefighters into the building." I held my microphone near his face. It was at this point that the chief changed his mind. The firefighters were obviously disappointed with the decision; we blamed command for not letting us have any fun. Firefighters and officers will fight fire the way we train them.

What we want them to do at real building fires is vent the building and flow water; do not risk firefighters to save unsalvageable property. Though what we teach them to do is take a beating, burn your helmet, see how close you can get to the fire, get in there and get it. All very macho and very aggressive but in all actuality, this is not very smart or safe.

As Chief Alan Brunacini said, "The hardest thing to do is put a firefighter in reverse." That is because we only teach them to go forward. The strategy and tactics of hesitation, retreat, wait for water, and wait for ventilation are not taught. Navy SEALs are taught tactical retreat, so it must be a macho decision under some conditions. Chief Ted Golden of the Mokena, Illinois, Fire Department asked these questions at a recent RIT/Mayday training session. 'Would you put a three-hundred-thousand-dollar firetruck in the building to attack the fire? If the answer is no, then why would you put firefighters in the building to attack the fire?" Very thought-provoking and soul-searching questions.

The purpose of much of our live-fire training seems to be, Let's see how hot and smoky we can make it for them so they become tough, aggressive firefighters. Maybe we need to rethink the objectives of live-fire training. At a minimum, live-fire training should be conducted as engine and truck operations all the time; ventilate to remove heat, flammable gas, and smoke; apply water to cool and extinguish. If it is hot and smoky maybe you forgot to do something. All of us from rookie to fire chief need the training: to teach the newbies how to do it right the first time, to teach the oldies to change their ways and finally get it right. When we do this, the mantra "Train the way you fight—fight the way you train" will save firefighters.

The Tale of Two Proactive Fire Chiefs
2008

IN 1977 DISTRICT of Columbia Fire Chief Burton Johnson invited the citizens of the city to visit any fire station to learn all they needed to know about smoke detectors. In 2007 D.C. Fire and EMS Chief of the District of Columbia Dennis Rubin launched a program to have firefighters visit every home in the city to ensure that families have working smoke detectors. "It was the best of times. . ."

In the '70s Washington, D.C., had some of the highest fire death rates in the nation (49.9 per million population). "It was the worst of times. . ."

During the '70s residential smoke detectors became available. The District of Columbia was one of the first metro cities to pass smoke-detector legislation for new and existing residential occupancies to reduce devastating fire death rates.

Every D.C. firefighter was trained on residential smoke detectors in 1976. This training included why smoke detectors were needed, how they worked, how they were installed, and what the codes, standards, and laws were. In addition, a smoke detector training manual was developed and distributed to every company in the department.

A mass media campaign called "Operation Return" was initiated to capitalize on the public's moment of interest when fire tragedy occurred. Community fire-safety education personnel would return to the location of fatal fires and explain to the media what would have prevented the tragedy. In most cases, a smoke detector would have given the occupants and victims time to escape. Local meetings were held to help the community learn from the tragedy to keep it from happening again.

Thirty years of smoke alarms have paid fire safety dividends. "It was the best of times. . ."

In 2006 Washington, D.C., had twelve fire fatalities. But one fire death is too many if it is your child—especially if the smoke alarm was not working.

On the night April 28, 2007, a fire started in five-year-old Asia Sutton's home. Unknown to her family, the smoke alarms were not working. When Asia's mom and dad woke up it was too late for them to rescue their daughter, and firefighters had no chance to save her life. "It was the worst of times. . ."

Unfortunately, this type of tragedy is repeated time and time again in our country. The U.S. Fire Administration estimates that as many as forty percent of residential smoke alarms today are not working. How many smoke alarms are not working in your community?

On July 21, 2007, Fire and EMS Chief Dennis Rubin along with Mayor Anthony Williams, the D.C. City Council and Mr. and Mrs. Sutton launched the Smoke Alarm Verification and Utilization (SAV-U) program. The Fire and EMS Department will visit every home to make sure the smoke detectors are working, replace batteries, and install new detectors if needed. If the family has young children the department will supply smoke alarms that can be programmed with the voice of the child's mother. Research indicates that mom's voice is the most effective sound to wake up sleeping children.

On the third Saturday of each month the District of Columbia will "Take it [SAV-U] to the streets. The D.C. Fire & EMS Department, City Council, mayor's office, local volunteer groups, IAFF Local 36, and corporate sponsors will go door-to-door to make sure every family has working smoke alarms. This program will continue until the citizens of D.C. are sure that all smoke alarms can SAV-U. "It will be the best of times!"

If the nation's fire service ensures that all smoke alarms are working, we can all say, "It is a far, far better thing that I do..." That tale will be a great story.

The White Elephant: Fire Department Response Time
2014

> A white elephant is an idiom for a valuable but burdensome possession of which its owner cannot dispose and whose cost (particularly cost of upkeep) is out of proportion to its usefulness or worth. The term derives from the story that the kings of Siam were accustomed to make a present of one of these animals to courtiers who had rendered themselves obnoxious, in order to ruin the recipient by the cost of its maintenance. –Wikipedia

FIRE OFFICERS TELL me that fire department management, city managers, and elected officials hold them accountable for getting out of the station in sixty seconds and want to know why they do not meet that standard. NFPA 1710 has a four-minute arrival time for the first engine ninety percent of the time and eight-minute arrival time for the full alarm ninety percent of the time. I do not know of any fire department that meets that standard. But every fire department and city reports the response time and it is a measure of a job well done even if the occupants are injured or killed. Response time is used to keep a fire station open or justify closing it. We never get there fast enough when your house is on fire.

Society is addicted to the notion of response time because we are stuck on the manual fire suppression model of fire safety. In other words, when a fire starts we expect a group of people to come rescue the occupants, put the fire out, and keep the fire from spreading and destroying the community. This construct is left over from Ben Franklin's notion of fire protection.

The manual fire suppression model is still the most prevalent fire protection model today and will be for the foreseeable future. It explains why thirty-two states have passed laws forbidding mandatory state residential fire-sprinkler building codes for new construction. The stated rationale is the added cost of fire sprinklers to new construction is not worth the money and will negatively impact home construction and sales. (Even though the cost of home fire sprinklers today is less than two dollars per square foot for new construction.) This idea rejecting home fire sprinklers is driven by homebuilders associations, and the fire service has not been able to counteract this political influence. So society will continue to use the response time of firetrucks as its measure of fire safety.

Put these in the order of importance. How fast the firetruck arrives at the incident. How fast the firetruck leaves the station. How fast dispatch sends out the alarm. How fast 911 notifies the fire department. How fast someone notifies 911 that there is a fire. How fast the fire is discovered, occupants decide to get out of the building and call for help.

If you do not buy the white-elephant argument let's try science. Flashover can happen in three minutes. That is before anyone can discover the fire, escape the building, call 911, dispatch the FD, and we respond. If you do not have smoke alarms and fire sprinklers we cannot save your family and your property.

Does the fire service have the courage to tell the truth about response time? A fire department's average response time is less important than the number of homes that have working smoke alarms. The fire service can never have enough fire stations, firetrucks, and firefighters to take the place of residential fire sprinklers. Please put these numbers and facts in your next annual report. If we don't, we will continue to keep the fire service response-time white elephant to the determent of public fire safety and firefighter safety.

When the king tries to give you a white elephant, don't take it. Ask for smoke alarms and residential fire sprinklers because that is the science behind measuring fire safety response time. No matter where you live, even in Siam.

Higher Education

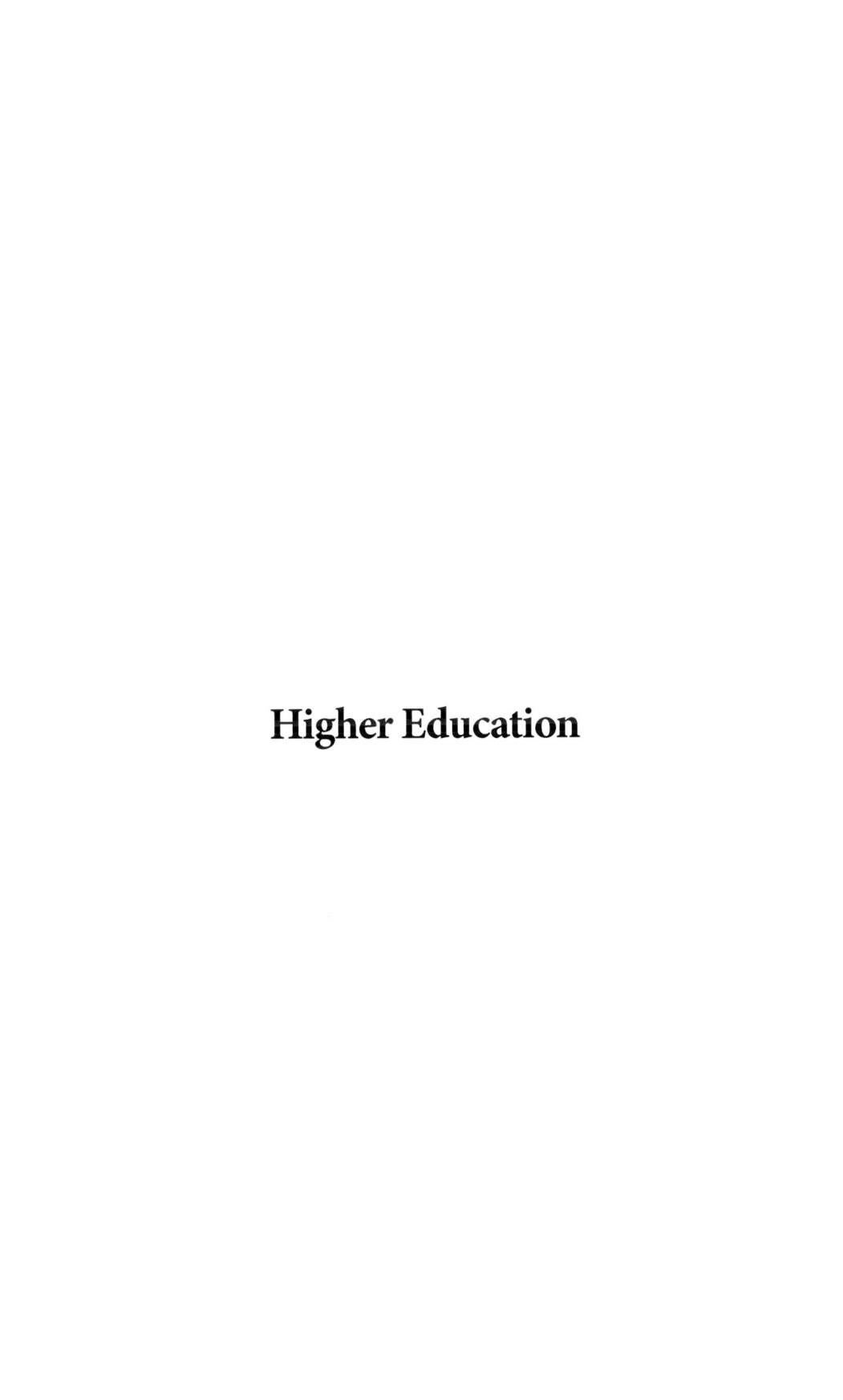

Feel the Door
1982

WHEN WAS THE last time you taught someone to "Feel the door; if it's hot, don't open it"? Did you ever have to explain what you meant by hot? A recent study by James Feld and Burton Clark, titled "Fire Study on Door Surface Temperature as an Indicator of Exit Tenability," attacks the validity of this instruction.

In the introduction to the study the authors bring up three questions.

1. Can an individual feel a door and determine if it's safe to open it?

2. What's safe?

3. Do you feel with the back or front of the hand, fingertips, or palm?

To answer these questions, Feld and Clark reviewed the literature and conducted a full-scale fire study.

They first addressed the question "What's safe?" They used 150 degrees Fahrenheit at five feet above the floor in the exit path as the temperature that indicated unsafe or untenable. Their references for this are "Detector Sensitivity and

Siting Requirements for Dwellings," NFPA 1976, p. D6, and "Operation School Burning," NFPA 1959, p. 25.

The full-scale study used a mobile home, 45'7"x18'x7', four bedrooms, kitchen, bath, living room, metal skin over wood frame, wood-paneled walls, fiberboard ceiling, and insulated. The bedroom door used was hollow-core, wood, 78"x30"x1³⁄₈", ³⁄₃₂" plywood front and back with a metal door knob. They started the fire in the kitchen using fifty pounds of ordinary combustibles and a match. Temperature data were read on a ten-channel digital thermometer with type-K thermocouple wires. The following table shows the locations of the nine thermocouples and the temperature reading during the five-minute burn.

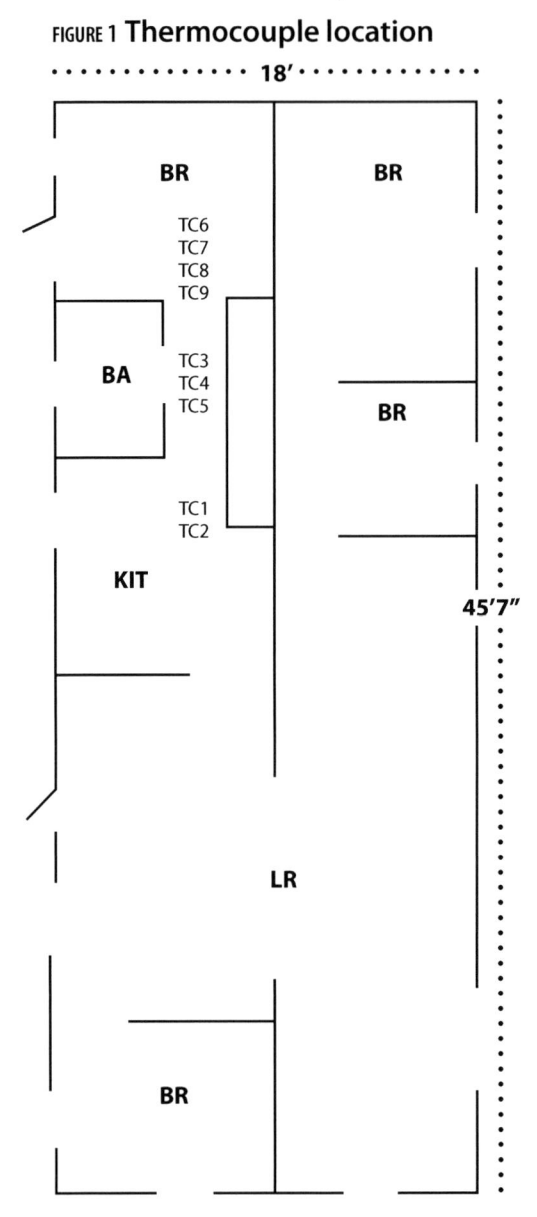

FIGURE 1 **Thermocouple location**

TABLE 1 Temperature data (Fahrenheit)

Minutes into Test	KITCHEN		HALLWAY			DOOR			KNOB
	TC1 4" from Ceiling	TC2 5' from Floor	TC3 4" from ceiling	TC4 5' from Floor	TC5 1' from Floor	TC6 4" from Ceiling	TC7 5' from Floor	TC8 39" from Floor	TC9 39" from Floor
1	316	241	246	154*	72	81	77	77	77
2	383	286	343	252	73	88	81	77	77
3	329	246	302	234	77	100	84	79	77
4	513	315	369	257	79	109	88	79	79
5	94.8	705	748	365	90	117	91	81	79
6	1454	1292	842	759	122	165	111	90	82

*The exit path was blocked one minute into the study.

Table I shows the temperatures at the nine different locations in the building during the six-minute burn period.

To answer Questions 1 and 3 they needed to determine the temperature of the hand. To do this they measured the hand temperature of ten adults at three locations on the hands. The average hand temperature was 90.03°F.

TABLE 2 Hand temperature data (Centigrade)

	MALE				FEMALE		
Subject	Finger	Palm	Back of Hand	Subject	Finger	Palm	Back of Hand
1	34.3	34.3	33.2	6	23.6	27.5	29.3
2	28.2	31.5	31.1	7	32.6	34.8	33.2
3	34.0	33.0	33.3	8	31.2	32.9	31.4
4	32.9	33.2	32.2	9	33.8	34.9	33.8
5	31.9	32.3	31.6	10	33.9	34.5	32.8
Average	32.26	32.86	32.28		31.02	32.94	32.1

Average Male: 32.47   Average Female: 32.02

Average Adult: 32.24°C = 90.03°F

Table 2 shows the various hand temperatures of the ten subjects.

In terms of question 3, the study states: "As may be seen

from the data in Table 2 there is not much difference between the temperatures of the various parts of the hand, because of this, this study does not address the question of which part of the hand should be used to feel a door."

In their discussion, the s explain the concept of contact temperature. The important point is that a door will feel cool as long as the hand temperature is greater than the door temperature. "Thus, until the door temperature rises above hand temperature (average hand temperature 90°F, it will feel cool."

Table 3 graphically shows the problem of teaching "if the door is hot. This graph shows the relationships between door temperatures in comparison to average hand temperature.

TABLE 3 **Bedroom door surface temperatures** (Unexposed side)

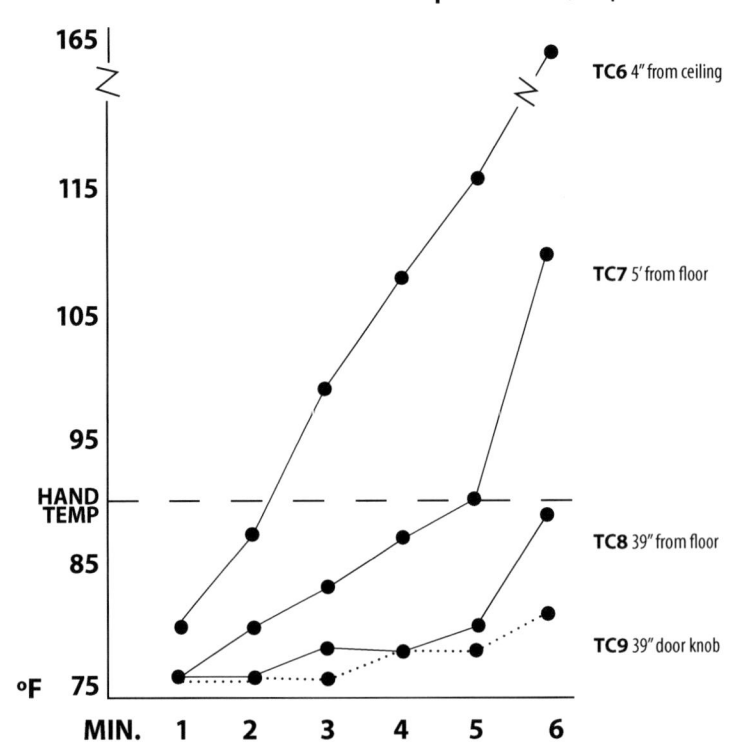

The conclusion states: "No absolute conclusion can be made from the data presented in this study. But under the given study situation and the present methodology of feeling a door, a person would not have been able to determine if it was safe to open the bedroom door by feeling it.

The words *hot* and *warm* are subjective. How "warm" or "hot" must a door feel in order for a person to make the decision not to open the door? New terminology and new methods need to be developed in terms of teaching people when to open the door. The instruction "Feel the door; if it's hot..." needs to be re-examined. If we have discovered anything it is the need for more study.

The authors hypothesize an educational suggestion.

"If a person feels a door and it's above hand temperature, the door should *not* be opened. One method to teach this concept is to have students feel each other's hand. If the partner's hand is above their hand temperature the student's response should be, "If you were a door I would *not* open you. I would use a second exit."

The problem still remains that a blocked exit can exist before the door surface temperature goes above hand temperature. It may be possible to feel the door to detect a difference in temperature. By feeling the bottom of the door, the middle of the door, then the top of the door, it may be possible to detect a difference in temperature, which would indicate the need to use a second exit.

This article was co-authored by James Feld.

Higher Education and Fire-Service Professionalism
1993

AS A FIRE CHIEF, you represent the pinnacle of fire-service professionalism, but your five bugles are only as strong as the higher-education infrastructure that supports them. If the fire service is to survive and prosper in the next century, fire chiefs must lead the way by recognizing higher education's importance for their profession and for themselves.

The importance of higher education to the fire service is not new. In 1966, at the first Wingspread conference, fire-service leaders stated that "professional status begins with education." They explained that higher education identifies a systematic body of knowledge, sets standards of conduct, helps the advancement and dissemination of knowledge, and identifies minimum standards.

At the first Wingspread conference, the means of achieving professional education were mapped out in a three-tiered educational model. The first step was for firefighters from probationary to craftsman, who would be trained and educated through extension courses and vocational schools. At the second step, junior officers and technicians were trained at the associate-degree level by technical institutions. Finally, chief officers and fire executives would go to baccalaureate programs at universities.

Wingspread I fire service education model, 1966

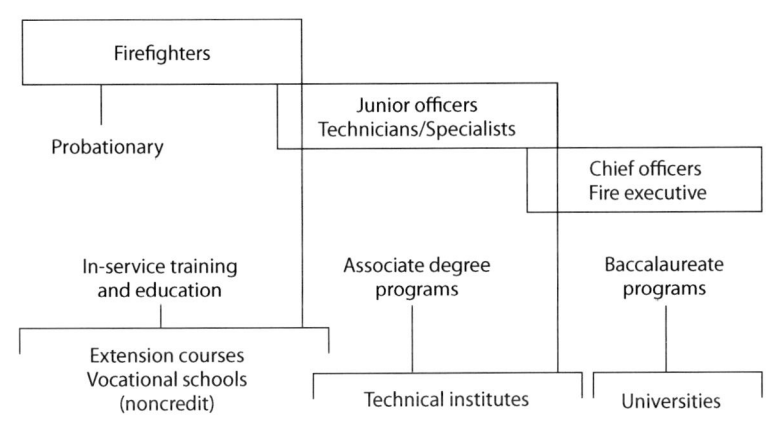

Firefighters (Probationary)	In-service training and education / Extension courses, Vocational schools (noncredit)
Junior officers, Technicians/Specialists	Associate degree programs / Technical institutes
Chief officers, Fire executive	Baccalaureate programs / Universities

The Wingspread group identified seven content areas in which the academic community could help meet the fire service's educational needs:

- First, mastery of the scientific method, because true professions are based on scientifically sound theoretical and empirical foundations.
- Second, an understanding of human relations, which leads to
- Content areas three and four, effective communications and organizational skills.
- The ability to concentrate while still maintaining an open and flexible mind were the fifth and sixth areas identified.
- Finally, the conference noted, professionals have to keep learning, on and off the job.

The educational model and content areas were identified because the leaders believed that "A systematic and deliberate educational program leading to a broad knowledge base which is acceptable to the academic community is the surest approach to professionalization."

In 1976, Wingspread II evaluated fire-service progress in higher education this way: "The development of fire service educational programs in the U.S. over the past ten years has been non-systematic and nondirectional. It is the responsibility of the educational community to sensitize itself to the needs and prepare students systematically to meet those needs."

Wingspread II also developed an educational model. The two major changes were the inclusion of a private-sector path for the fire protection engineering community and the inclusion of graduate school education at both the master's and doctoral degree levels.

Wingspread II fire service education model, 1976

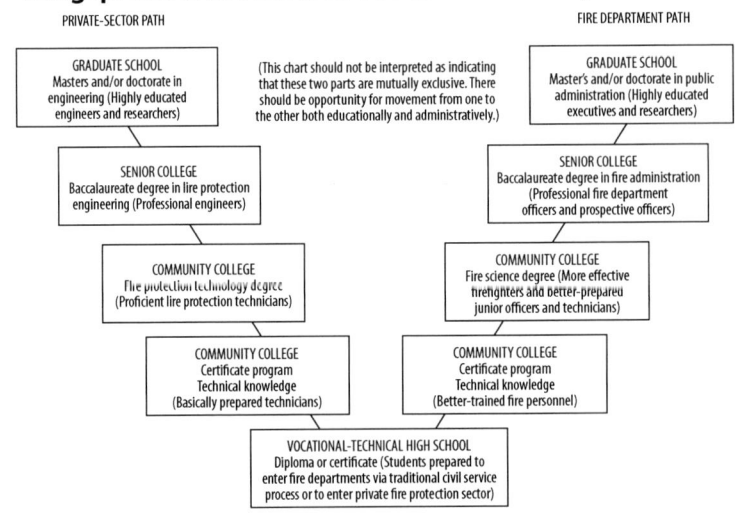

PRIVATE-SECTOR PATH | FIRE DEPARTMENT PATH

GRADUATE SCHOOL
Masters and/or doctorate in engineering (Highly educated engineers and researchers)

(This chart should not be interpreted as indicating that these two parts are mutually exclusive. There should be opportunity for movement from one to the other both educationally and administratively.)

GRADUATE SCHOOL
Master's and/or doctorate in public administration (Highly educated executives and researchers)

SENIOR COLLEGE
Baccalaureate degree in fire protection engineering (Professional engineers)

SENIOR COLLEGE
Baccalaureate degree in fire administration (Professional fire department officers and prospective officers)

COMMUNITY COLLEGE
Fire protection technology degree (Proficient fire protection technicians)

COMMUNITY COLLEGE
Fire science degree (More effective firefighters and better-prepared junior officers and technicians)

COMMUNITY COLLEGE
Certificate program
Technical knowledge
(Basically prepared technicians)

COMMUNITY COLLEGE
Certificate program
Technical knowledge
(Better-trained fire personnel)

VOCATIONAL-TECHNICAL HIGH SCHOOL
Diploma or certificate (Students prepared to enter fire departments via traditional civil service process or to enter private fire protection sector)

Another important aspect of this model is the identification of a discrete discipline, or content area, at each academic level. At the community-college level, fire-protection technology and fire science are listed. At the senior-college level, baccalaureate degrees in fire-protection engineering and fire administration are listed. Finally, at the graduate-school level, the concept of fire is dropped out, and engineering and public administration are identified.

The most recent conference, Wingspread III in 1986, did not address higher education. The only relevant statement made was "professional development in the fire service has made significant strides, but improvement is still needed."

Fire vs. police education. The attendees at Wingspread I and II clearly understood the importance of higher education if the fire service is ever to achieve professional status. To illustrate the correlation of higher education to professionalism, we can compare the fire service to the police service.

There are more than 1,100 associate-degree programs related to the police service, compared to 314 fire service programs. At the baccalaureate degree level, there are thirty-six fire programs and 648 police programs. At the master's degree level, there are four fire and 130 police programs. Finally, at the doctoral level there are nineteen police programs and just one fire program.

Police vs. fire: Number of degree programs offered

Discipline	Associate	Bachelor	Master	Doctorate
Law enforcement/criminology	1,141	648	130	19
Fire science	314	28		
Fire protection engineering	8	4	1	

Higher education for the police service clearly outweighs the fire service at all levels. But the most striking difference is at the graduate school level, where the fire service as a discipline is almost nonexistent. Even the Wingspread I and II groups did not consider fire science a discipline to be studied at the graduate school level.

To the extent that law enforcement is more of a profession than the fire service, it is so largely because of its inherently close association with the legal profession.

Similarly, EMS providers, both within and outside the fire service, are driven to greater professionalism through constant contact with highly educated medical professionals.

The importance of postgraduate education to a profession is clear when the purpose of such programs is understood. Lucht describes the purpose of such programs as follows:

> Master's programs, offering graduate degrees in the discrete discipline, training practitioners for the job market and in preparation for doctoral work. Graduate students also serve as a pool of workers to help professors with their teaching and research, contributing to the body of knowledge.
>
> Doctoral programs, offering degrees in the discrete discipline, preparing highly specialized expertise for industry and creating a pipeline of qualified personnel to serve as faculty; doctoral students also help professors with their teaching and research, making major contributions to the body of knowledge.
>
> College and university faculty, with doctoral degrees in the discrete discipline, teaching future practitioners, driving important research to add to the body of knowledge, and writing definitive textbooks.

The three central concepts associated with graduate schools and degrees are discrete discipline, research, and body of knowledge.

Discrete discipline: When an area or subject matter content is identified and used almost exclusively by practitioners of that subject area, it becomes a discrete discipline. For example, accounting, engineering, law, and medicine are discrete disciplines.

The argument can be made that the fire service does not constitute a discrete discipline, because it uses other disciplines to practice its art and science. The fire service uses engineering, law, management, medicine, education, political science, chemistry, and physics, all of which are themselves discrete disciplines. The fire service practices these disciplines under unusual conditions and in unique environments, however, which in many cases changes the fundamental theoretical foundations of these disciplines, or at the very least affects the empirical research results.

For example, the fire service is the only occupation that regularly employs personnel on a twenty-four-hour shift and has them sleep on the job. Sleeping on the job does not fit into standard economic or personnel formulas. Neither does the fact that about eighty percent of fire-suppression service is conducted by volunteer personnel.

Even the concepts of chemistry and physics change when removed from the controlled environment of the laboratory to the uncontrolled environment of emergency operations.

Research: If you don't accept the argument that the fire service is a discrete discipline, you must realize how lacking fire-service research is. Graduate schools conduct and sponsor the research that builds the body of knowledge the practitioner uses.

Again, we can compare fire and police. In 1992 there were 196 doctoral dissertations that related to the police service, but only five dealing with the fire service were conducted. *Dissertation Abstracts*, the national database for doctoral research, does not identify fire as one of its 248 subject areas, though criminology and penology are listed.

(Some of the more esoteric subjects include cinema, home economics, and folklore. There are nine master's degree programs and five doctoral programs in folklore, and eighty-five dissertations were classified under folklore in 1992.)

The importance of research to the fire service was

acknowledged by the Wingspread II group when it identified one of the products of graduate schools to be highly educated researchers. Wingspread I identified the "mastery of the scientific method" as the fire service's first educational need.

Body of knowledge: A discipline's scientific research is what generates its body of knowledge. The body of knowledge for the fire service can be found in its literature: books, journals, manuals, standards, and educational and training materials.

Much of the fire service's body of knowledge is based on consensus and experience, neither of which is considered a scientifically sound methodology.

The quantity and quality of the literature a discipline is based on also serves as a measure of its professionalism. A comparison of fire and police periodicals illustrates the difference.

Ulrich's periodicals directory lists 220 periodicals under criminology and law enforcement published in the United States, compared to fifty-five under fire prevention. In addition, only one fire periodical was listed as being a "refereed serial," compared to twenty refereed serials for criminology and law enforcement. Refereed serials (journals) are considered more scientifically sound, because the material is reviewed and approved by a panel of judges before publication.

Higher education and you: Professionalism is complicated, but the fire service's becoming a professional occupation doesn't mean that every fire chief needs a PhD. Few police chiefs have doctorates.

The fire-service profession's needs are identical to what Lucht said is needed for the fire protection engineering profession: "A permanent academic infrastructure must be put in place to train practitioners, perform research, and produce advanced scholars of the next generation. This includes strengthening existing programs and expanding the family of curricula at the BS, MS, and PhD level."

So what does all this have to do with you? Presumably your professional development has not stopped because you have five bugles. Many of you already have bachelor's degrees, and some of you hold master's degrees. If an opportunity for continued graduate study and research were made available, some of you would take advantage of it.

Once a fire-science academic infrastructure is in place, more people will choose teaching and research as their career path. I believe there is enough funding from the insurance, apparatus, and equipment industries and from other philanthropic sources to support fire-service higher education.

If you don't want to go to school any more, and doing research does not sound like fun, the results will still be valuable to you. If you had scientific research to show why you need an automatic defibrillator on each engine, your chances of getting them would be greatly increased. If we investigated each firefighter death as completely as we investigate each airplane crash, the research would save lives.

Finally, these concepts apply to you because fire chiefs are the cultural leaders of the fire service. As a discipline, in the twentieth century, we have been training-oriented, from the bottom up, with an experiential/consensus knowledge base. What we need to become, for the twenty-first century, is education-oriented, from the top down, with a research-science knowledge base.

If you, the fire-service leaders of today, believe and say that higher education and research are important, they will become important and the systems will be created. In the meantime, maybe your next firefighter of the year award will go to a member who has conducted an outstanding research project.

In 1861, Yale University awarded the first PhD degrees in this country to Eugene Schuyles, James Morris Whiton, and Arthur Williams Wright. In the year 2000, who will be the first students to receive their PhDs in fire science, and from what school will they graduate?

References

Bowker, R.R., Utrich's International Periodicals Directory, Vol. 1. (Reed Publishing Co.; New Providence, N.J.: 1992).

Clark. William E., Wingspread II: Statements of National Significance to the Fire Problem in the United States (International Association of Fire Chiefs Foundation Inc.; Washington, D.C.: 1976).

Fire Service Institute/Iowa State University and The Johnson Foundation, Wingspread III Conference on Contemporary Fire Service Problems. Issues and Recommendations (International Association of Fire Chiefs Foundation Inc.; Washington. D.C.; 1986).

Johnson Foundation, Wingspread Conference on Fire Service Administration, Education and Research (The Johnson Foundation; Racine. Wis.; 1966).

Lucht, D.A. "Coming of age." *Journal of Fire Protection Engineering*. Vol. 1. No. 2, pp. 35-48, 1989.

Peterson's Guide to Four-year Colleges 1992 (Peterson's Guides: Princeton, NA.).

Peterson's Guide to Graduate and Professional Programs. An overview 1992 (Peterson's Guides: Princeton, N.J.).

Reading and Writing Equal Professionalism
1999

WHAT READING AND writing skill levels do fire chiefs need to attain? According to the U.S. Department of Labor (1991) a fire chief's language skill level should be the same as a physician. The skills include reading scientific and technical journals, writing editorials, journals, speeches, and manuals, and critiquing and speaking at the discussion and debate level. This high level of reading and writing at the scientific level will come as a shock to most of you. But in 1966, at the first Wingspread conference, fire service leaders of the day identified mastery of the scientific method as the first educational need of the fire service for it to become a true profession.

Today the fire service is becoming a writing discipline. In the NFPA Fire Fighter I and II professional qualification standards (1997), writing skills are identified fifteen times. "Communicate in writing" is listed forty times as a prerequisite skill in the Fire Officer I, II, III, and IV standards (NFPA, 1997).

The Maryland Fire and Rescue Institute Fire Officer III curriculum (1998) requires the student to conduct graduate-level applied research. To graduate from the National Fire Academy Executive Fire Officer Program (2001) students must successfully complete four applied research projects at the graduate school level of performance. We must include reading and writing throughout our training, education, and practice from entry level to top scholar/practitioner to achieve these standards and increase our professionalism.

Every basic firefighter school needs to have a writing section as part of the curriculum. Every rookie police school already does, so guidelines are available and can be modified as needed. Look for computer-based training programs that are language arts tutorials. Make rookies keep journals of their first-year experiences. Have company and chief officers review their journals.

At the associate's degree level, students should be writing synopsis of fire/rescue service magazine articles. Then they should be taught to compare and contrast several authors who write on the same topic. Make sure they use proper citation methods. When firefighters and fire officers get to the junior and senior level of bachelor programs they should be reviewing articles from refereed journals. To do this they will have to read outside the fire-service discipline since we do not have any refereed journals. By the time we achieve a bachelor's degree we should be able to write a comprehensive literature review on any topic assigned. This will prepare us for scientific study and eventually conducting and critiquing our professional research.

You may think comparing fire chiefs to physicians in terms of reading and writing skills is an intellectual stretch. But Chief Massy Shaw of the London Fire Brigade did just that in 1873 when he compared the training and education levels needed by a fire chief to that of a surgeon (Layman, 1953). Chief Shaw would think we are headed in the right direction with our training, education, and standards. but after a century he would wonder what has taken us so long to see the light.

References

Johnson Foundation. (1966). Wingspread conference on fire service administration, education and research. Racine, WI: Author.

Layman, L. (1953). Fire fighting tactics. Boston, MA: National Fire Protection Association.

Maryland Fire and Rescue Institute. (1998). Fire officer III. College Park, MD: Author.

National Fire Academy. (2001). Executive fire officer program guidelines. Emmitsburg, MD: Author.

National Fire Protection Association. (1997). Fire fighter I, II professional qualification standard 1001. Quincy, MA: Author.

National Fire Protection Association. (1997). Fire officer I, II, III, IV professional qualification standard 1021. Quincy, MA: Author.

US Department of Labor. (1991). Dictionary of occupation titles. Washington, DC: US Printing Office.

August 7, 2001

The Best of the Best
1999

EACH YEAR, A mere handful of Executive Fire Officer Program students at the National Fire Academy receive the Outstanding Research Award. Here's what it means to win this exceptional honor.

What do Dawn Smith, J. Curtis Varone, Thomas French, and Robert B. Laughlin have in common, and why should you care?

Answer: They're all winners. Laughlin received a Nobel Prize for physics, French was awarded a Pulitzer Prize for feature writing, and Smith and Varone received Outstanding Research Awards for their applied research projects completed as part of the National Fire Academy's Executive Fire Officer Program.

You might not be interested in physics or journalism, but as a fire chief you should be reading papers titled "Correlation of a Physical Fitness Evaluation Test to a Selection of Firefighting Tasks" or "Fireground Radio Communications and Firefighter Safety" because either or both might affect you and your organization.

What's the impact of receiving an ORA on these individuals and their departments? That's the question I tried to answer in my own ARP, on which this article is based.

A history of excellence

The National Fire Academy describes the purpose of the EFOP ARP ORA as follows: "The Nobel Prize, the Pulitzer Prize, and the Olympic Gold Medal are recognized worldwide as symbols of human excellence. These awards are bestowed on individuals in recognition of achievements that have made significant contributions to society. . . .

"A significant measure of the professionalism of any occupation is the quality and quantity of the research the discipline is based on. The Outstanding Research Award helps the NFA achieve its mission of increasing the fire service's professionalism."

Since 1989, over four thousand EFOP ARPs have been written, all of which are archived in the National Emergency Training Center's Learning Resource Center, in Emmitsburg, Md. The research papers are graded on a zero to four scale. A 4.0 equals a letter grade of A. The evaluators are NFA adjunct faculty who have extensive fire service and academic backgrounds.

Over the past ten years, 415 ARPs from forty-two states plus Australia and Canada, have been graded 4.0. These papers are listed in the EFOP ARP selected bibliography, which is available on the USFA's website (www.usfa.fema.gov). The ten states claiming the most 4.0 research papers are California (52), Florida (41), Texas (36), Washington (21), Colorado (20), Ohio (20), Maryland (15), Virginia (15), Arizona (11), and Illinois (11).

The NFA faculty selects the ORA winners from the A papers written in a given calendar year. In the past ten years, only forty-two ARPs have received the award. In other words, only about one in ten ARPs receives a 4.0, and only about one in ten of those is selected for the Outstanding Research Award. Thirty-two EFOP students from around the country have been honored for their contributions to fire-service professionalism through their research. Six of these researchers have received the award twice or more.

Winning papers are published annually by the USFA; they're also available on the USFA website. A display honoring the award recipients was created in the academy's classroom building, with the winners' pictures, organizations, and research titles. The winners are also invited to present their papers at the EFOP Graduate Symposium held annually at the NFA, where they receive their award certificates. In addition, in 1990, the Society of Executive Fire Officers began presenting a cash prize to the award recipients.

The research approach

To gather the information on the impact of receiving an NFA ORA, I sent a letter to each of the twenty-five winners from 1989 to 1995. In the letter I asked them to identify the personal, professional, and organizational impact of receiving the award. Twenty-four winners wrote back.

The letters were considered the raw data, which I reviewed using a process called thematic analysis to identify themes or patterns in the data. Once the patterns in each letter were identified, all the letters were combined to identify the themes being reported for each impact type (personal, professional, and organizational).

The next step was to classify each theme by one of three outcome classifications: material consequences, social reaction, and self-reaction. Each theme was also classified as being positive or negative. The operational definitions used for these concepts were:

Material consequences: Anything that can be associated with money, job, time, work, goodwill, a tangible item, written articles, award, and reward.

Social reaction: Principally associated with others' feelings, beliefs, values, and judgments.

Self-reaction: Principally associated with one's own feelings, beliefs, values, and judgments.

The final step was to group the principal themes together and call the result the impacts. A major limitation to the study is the fact that I was the only one to review, identify, and classify the data, themes, and impacts. It must be kept in mind that the process is subjective, based on my knowledge, beliefs, values, and judgment.

The results, please

The results included 158 themes, the majority of which were classified as being positive, although twenty-one were negative. The majority of the negative themes came under the organizational impact type, such as: news reporting being hit and miss, research not used to benefit citizens, increased union/labor tension, and research never recognized. The overwhelming majority of the themes, however, were positive.

The type of impact most frequently reported by the EFO laureates (eight instances) was personal. The top personal impact for self-reaction was self-confidence; for social reaction, pride; and for material consequence, professionalism.

Six impacts were identified under the professional-impact type. The top impact for self-reaction was professionalism; for social reaction, recognition; and for material consequence, opportunity. Three impacts were identified under the organizational-impact type. There was no self-reaction impact identified. Pride was the social reaction, and reputation was the top material consequence. The outcome classification with the most impacts was material consequence (eight instances). Self reaction was second with five, and social reaction was third with four.

To get a sense of what the winners wrote, I identified

quotations that captured the essence of the impacts. The laureates' own words explain the impact of receiving an NFA ORA.

Personal impact

Self-reaction. The sense of **self-confidence** that EFOP laureates feel can be very specific, as with Richard Arwood, who said the award gave him a "new dose of self-confidence regarding my writing skills." Or the award can provide a whole new experience because it was the "first time in my life that I have been individually and publicly recognized for academic excellence" (Randy Reiswig). Finally, Tom Alexander, who keeps the certificates on his wall, said, "looking at them has a way of renewing my confidence."

Pride was reported by many laureates. An anonymous laureate said he felt "a great sense of professional pride." Tom Alexander explained his pride: "Twice in my life I did something that was judged to be better than anyone else who tried. Regardless of who else knows, or cares, I know. I will carry it with me forever."

David Harlow called the ORA a "great **honor**" and "The greatest honor I have received during my twenty-four years in the fire service" is how Tom Wood described it. Dawn Smith was "honored to have my comments held in such esteem."

Making the phone call informing recipients they've received the award is one of my greatest pleasures and a total **surprise** to recipients. Dawn Smith said, "My heart jumped, I felt flushed, and a bit shaky... the wow effect." When she received a second award, her response was "disbelief. You're kidding, right? Again!?"

Some recipients reported negative impacts. Keith Brown indicated he was "frightened that I would not be able to meet others' future expectations." Bill Sager said it was a "downer to realize that it wouldn't change the world." Dawn Smith reported insecurity, "was it [the research] really good enough for such recognition" and anxiety, "I had to present this thing?!?"

Social reaction. Pride is the impact most reported by laureates. Family, friends, and associates all share in the award. Mike Kuypers wrote, "My wife was also very proud of this accomplishment." David Harlow said, "My parents probably received the biggest impact from my receiving the award. They were so proud to be able to come up to the NFA to see the facility and go to the dinner."

The social group also gives laureates positive **reinforcement** by sharing the joy, happiness, and recognition. Kerry Koen said the award was "recognition of team work at home." Bernard Dyer said, "My wife was very happy to hear that I had received the Outstanding Research Award. Besides being very supportive in my career she knew that I had put a lot of effort into developing a quality research report and she was pleased that I was recognized for that work."

John Hawkins also talks about family: "My family, two sons of which are also firefighters, were elated and constantly commended me for the recognition." Finally, Ed Hayman brings the impact back to the NFA: "It is the sense of respect from other EFO participants that made receipt of the award even more gratifying."

Material consequence. The themes related to the personal impact/material consequences of receiving an Outstanding Research Award are professionalism and symbolism.

Professionalism is hard to define. Tom Wood summed up one important aspect when he said, "I am also proud to record the awards on my résumé." Bernard Dyer said it "helped enhance my professional reputation within my department." For Curt Grieve, it meant money in his pocket: "[I]t has assisted me numerous times in being selected over other fire-service consultants in today's competitive world." The award "confirmed I am doing high caliber and worthwhile work" (Bernie Williams).

The anonymous winner's statement captures the value of outstanding research: "The thought that others may use the research to benefit the safety and welfare of the people they serve was very rewarding." Lastly, many laureates also receive additional awards from their departments, local, and state governments, and professional organizations.

Symbols are important; they stand as constant reminders of what's important. The EFOP Outstanding Research Award photo display in J Building has come to symbolize fire-service excellence. Richard Arwood sums it up, "The photo on the wall is one of the greatest rewards to me."

Professional impact

Self-reaction. Again, the laureates express **professionalism** in different ways. Bill Sager said that after reading his paper, people "would contact me and validate the conclusions that I had developed." Tom Alexander describes it this way: "I believe [the awards] actually made me more professional . . . to be credible, I had to do my job in a manner befitting a national award winner." According to Mike Kuyper, there's "increased prestige both within and outside the organization that comes with the award." "A national award adds to one's professional credentials" (Kerry Koen).

Social reaction. Outstanding Research Award laureates receive **recognition** from other groups in many forms. Curt Grieve wrote, "Receiving the award not only brought a very visible amount of pride to the Board of Directors that I worked for but had the seeming effect of them displaying a greater . . . confidence in my ability."

The award brings attention to the research. Curt Varone: "[S]everal high-ranking city officials have read both of my papers and have complimented me on them. Had the papers not received the awards, I doubt anyone (whether from the department or the city administration) would have read them."

The recognition's value depends on how the laureate

views the group. Ed Hayman: "... the recognition and sincere congratulation of my fire-service friends throughout the country and state had the greatest impact. These are people I respect."

On the downside, it's disappointing when others don't care about your work. Richard Jioras: "[T]he award and the subject of the report passed unnoticed in our department.... There was little departmental or community recognition or impact as no one knew nor seemed to care that I received the award." And the Outstanding Research Award doesn't escape the green-eyed monster. Richard Arwood reported that "[Some of my peers make] teasing remarks indicative of the fact that they are envious of my honor."

Material consequence. For Greg Gentleman, there were two **opportunities**: "an immediate increase in visibility within the fire service and an increased recognition that a 'civilian' can actually have a positive impact...." For Bill Sager, the award "opened an opportunity to participate in course development at NFA and eventually to be an adjunct instructor." Finally, Leslie Bunte credits the award with his admission to graduate school: "For three years prior to the award I sought admission to graduate-level studies at the Lyndon B. Johnson School of Public Affairs at the University of Texas. Each year I was rejected for admission and was seriously on the verge of not re-submitting an application. However, on my next application after winning the [award], I was immediately selected for admission.... I probably would not ever have had the opportunity to obtain a Master of Public Affairs degree from this prestigious graduate school had I not received the Outstanding Research Award."

Tangible recognition by others brings credibility to the EFO Program and its laureates. Tom Wood, the Florida Fire Marshal's Association's Fire Marshal of 1996, said, "I am certain that my 'Outstanding Research Award' had some influence in being chosen." Steve Dalbey said, "During the two weeks prior to the [EFO] Symposium and awards banquet, I was at the Academy for my third EFO course. During that time I was 'summoned' to Senator Grassley's office to be congratulated."

The research and the award can also spark recognition from the private sector. Julius Halas: "The city of Sarasota and our Public Safety False Alarm Taskforce won subsequent recognition for the project, as presented by the National Electrical Manufacturers Association and... Operation Life Safety. Later, I was... honored by the National Automatic Fire Alarm Association as the 1993 Man of the Year for my contributions to false alarm reduction research."

Finally, Kerry Koen sums up community recognition: "It is a fairly basic... phenomenon: Almost every community likes to feel that its fire-rescue department is the 'best.'... National awards and other recognition give the community something tangible to point to to affirm these feelings."

Outstanding Research Award laureates receive **professional status**, which lasts over time. Seven years after receiving the award, Greg Gentleman still receives requests for his research. "It seemed to greatly increase my credibility, even on non-related issues. It gave me a host of additional contacts around the country and world from which to draw ideas and opinions in a wide variety of fire service issues."

Keith Brown believes the award helped secure his career position. "All three of my [job] interviewers were EFOP graduates and may have viewed my research award as being indicative of traits and capabilities equivalent to the educational qualification they were seeking...."

There is a connection between professionalism and publication. Craig Kampmier had his research published three times: in the NFPA newsletter, a biomedical newspaper, and the juried *Journal of Applied Fire Science*.

Do the research and award help foster **change**? They did in Sarasota, Florida, said Julius Halas: "[M]y research project evolved into the adoption and implementation of a comprehensive false-alarm ordinance within the city of Sarasota." Tom Wood said, "The accolades are wonderful; however, a deeper satisfaction comes from seeing ... my research accepted and implemented.... My department had sprinkler heads installed in all our fire station kitchens."

There can be a downside to successful research and high-profile recognition, as Curtis Varone learned: "One effect of receiving the award was to derail one of the City's options for downsizing the Fire Department. [M]y research [found] that staffing with four firefighters per apparatus significantly reduced the number and severity of injuries in Providence compared with three-person staffing. As a result of the notoriety given to the research by virtue of the award, the City of Providence was not in a position to advocate reducing staffing back to the three members as a means to save money. In this regard, it is conceivable (and some have made a point of telling me) that my chances for promotion may have been damaged...."

Organizational impact

Self-reaction. There was no identified theme among the laureates related to self-reaction to the organizational impact.

Social reaction. Curtis Grieve believes that the community expressed "their **pride** in the department ... and department members were excited about the award."

Bernie Williams said, "My chief was very impressed by the fact I received the award, and he forwarded copies of my paper to the city manager and the mayor." John Koehler asked his chief what he thought the award meant. "His remarks were that it has set a standard for other participants from our agency to meet or exceed."

The negative impacts are lack of recognition and jealousy (Kuyper, Harlow). For Curtis Varone, "the result of the research and subsequent awards has been to put me somewhat at odds with the firefighter union."

Material consequence. Outstanding Research Awards enhance the **reputation** of the laureate's organization. Dawn Smith said, "Local officials are aware of the awards and have at times mentioned them in public meetings in reference to the quality and progressiveness of the department."

This is true also for Bernie Williams: "The award helped [the department] to publicly gain . . . credibility, and had some influence over council members in persuading them that the department was in fact being professionally managed and that a full audit review was not needed."

Tom Wood said, "[M]y award and the many other honors bestowed on Boca Raton Fire Rescue Service confirm for the citizens of the community that they are getting the highest quality public service." Kerry Koen noted, "[E]very achievement helps focus on the quality of life in a particular community, and these achievements may . . . influence a corporation or individual to invest or reinvest in a community."

The award can serve as a good defense. Curtis Varone writes, "I doubt my research would have even been given any consideration whatsoever by politicians and the media (who had been merciless in their attacks upon the department). Since I received the second award, the public attacks on the department by politicians and the media have ceased."

Laureates report the direct **implementation of their research**. Bernard Dyer said, "[In] my target audience, the volunteer system in the counties surrounding Philadelphia, two communities decided to develop and implement written ICS guidelines for the volunteer fire companies. They asked me to help them implement a workable model and I also provided some training sessions."

In Providence, Rhode Island, the "results of my research make it unlikely that the city, or even an arbitrator, would agree to reduce staffing back to three members per apparatus" (Curtis Varone).

According to Julius Halas, the City of Sarasota, Florida,

Fire Department had a thirty percent reduction in false alarms the first year after the False Alarm Ordinance was implemented and ten percent to fifteen percent per year decreases thereafter. The ordinance is being extended to the entire county, an additional three hundred square miles.

The two most negative material consequences to the research and the award are no use and misuse. Dave Harlow wrote, "[T]he organizational culture study was used by the union in their ongoing fight against the fire chief. Based upon information found in the study, and things taken out of context, they held a unanimous vote of no confidence against the chief. They had no regrets in naming the source of their information as being from my research report."

Discussion and suggestions

"The Nobel Prize, the Pulitzer Prize, and the Olympic Gold Medal are recognized worldwide as symbols of human excellence." The EFOP ORA hasn't reached that level of recognition and presumably never will, but it is becoming recognized in the fire service as a symbol of outstanding contributions to our communities and our discipline.

EFOP students and their Applied Research Projects are contributing to the quality of life in their hometowns and organizations. You may not know a Nobel, Pulitzer, or Olympic winner, but there's probably an EFOP student, graduate, or laureate near you. Traditionally, the fire service has recognized heroic acts, fallen firefighters, and the busiest companies. The NFA EFOP ORA gives us an opportunity to spotlight accomplishments that we can all strive for and benefit from..

As fire chief, you probably don't need to keep up on the latest Nobel or Pulitzer winners, but you do need to review the work of NFA ORA winners or, at the very least, the abstracts of their work.

The study's most disappointing finding was the lack of

impact at the organizational level. If we as a discipline don't support, encourage, and recognize research and those who do it, we won't obtain the professional status we strive for. So, at your next county chiefs meeting, invite an EFOP student to present his or her research paper. At the state conference, recognize EFOP graduates and have an ORA candidate or winner present a paper.

Finally, an important function of leadership is to inspire others to achieve new heights of excellence. What do you present to your community and department as examples of fire-service excellence? I can't wait to see fire department T-shirts with the slogan "We're #1 in NFA Outstanding Research Awards."

What an impact that would be!

Who Needs a PhD?
2004

THE IDEA THAT the fire service needs to be studied, practiced, and researched at the doctoral level is not new. In 1868, Sir Eyre Massey Shaw, fire chief of the London Fire Brigade, visited major fire departments in the United States and made several observations, including the following:

"When I was last in America, it struck me very forcibly that, although most of the chiefs were intelligent and zealous in their work; not one that I met made even a pretension to the kind of professional knowledge which I consider so essential. Indeed, one went as far as to say that the only way to learn the business of a fireman was to go to fires—a statement about as monstrous and contrary to reason as if he had said that the only way to become a surgeon would be to commence cutting off limbs without any knowledge of anatomy or of the implements required."

Shaw's comparison of the fire service to the medical profession remains apt. Physicians practice medicine as an applied science. If there's no research to support a procedure, piece of equipment, medication, or diagnostic rubric, it is malpractice to employ it. The fire service is an applied-science discipline, so our practice should have its foundation in research.

The need for advanced academic study and research in the fire service was reinforced at the first Wingspread conference in 1966.

The academic ladder

But what have we done over nearly four decades to achieve that vision? Our path to achieving professional status has been from the bottom of the academic/research ladder.

We have associate and baccalaureate programs in fire-service concentrations. More recently some master's degree programs have emerged that focus on the fire service. In addition, the National Board on Fire Service Professional Qualifications System, the National Fire Academy Executive Fire Officer Program, and the Chief Fire Officer Designation program have contributed professional credentialing methods to our progress. These excellent efforts have taken decades to implement. Nonetheless, they fall short of the top rung of the academic/research ladder: the doctorate.

The doctoral degree/research academic infrastructure is what creates "a knowledge base [for a discipline] that is acceptable to the academic community," according to David Lucht. He notes that doctoral programs offer degrees in the discrete discipline, prepare highly specialized expertise for industry, and create a pipeline of qualified personnel to serve as faculty and doctoral students to help professors with their teaching and research, making major contributions to the body of knowledge. College and university faculty with doctoral degrees in the discipline teach future practitioners, who conduct important research to add to the body of knowledge and write definitive textbooks.

Once a doctorate is available, the occupation begins to achieve professional status because knowledge and practice can be based on science and research. Currently most fire-service knowledge is based on experience and consensus, neither of which is acceptable to academic and professional communities.

Some fire chiefs may respond, "Who cares if the academic community accepts what we do?" You may not need a college degree to ride the firetruck, but you do need scientific research acceptable at the highest academic and professional levels to justify the existence and cost of the firetruck, to determine how many personnel ride on the firetruck, and to measure the performance of the firefighters on the firetruck.

Police vs. fire academia

The academic community is the mechanism by which professional disciplines justify, determine, and measure the efficiency and effectiveness of their doctrine.

Examining the academic infrastructure available to the police will illustrate the shortfalls of the fire-service discipline. This disparity is a principal reason for the difference in economic and political support that the two primary public emergency services receive.

For example, *Peterson's Guide to Graduate and Professional Programs* reveals that twenty-nine universities offer doctorates in criminal justice or criminology. The National Science Foundation reports sixty-two research doctorates in criminology were awarded in 2001 in the United States. There are no doctorates in the fire-service discipline listed. There are four in fire-protection engineering, but that is not the fire service.

According to *Dissertation Abstracts 2003*, in 2002 seven fire-service-related doctoral dissertations were conducted, compared to thirty-four police service doctoral dissertations. A review by research assistant Adam Zile of the *Chronicle of Higher Education* for much of 2003 finds fifty-four college faculty positions advertised under criminology and just two under fire science. In addition, five new scholarly books related to criminology were reported, but none was listed for the fire service.

Even the Ivy League helps to give the police service professional status; Harvard University is part of the police service academic infrastructure. The Kennedy School of Government graduate program in Criminal Justice Policy and Management has twelve faculty, fellows, and researchers currently conducting seven major research projects related to the police service. There are no programs, faculty, or research related to the fire service listed at the Kennedy School.

Why they do it

Despite the fact that there is no fire-service studies doctoral program available, firefighters still are going for the gold tassel: the doctoral degree. Why do they do it, and what is the benefit to the fire service?

This question was posed to four fire officers who have earned doctorates: Ron Wakeham, past fire chief of Des Moines, Iowa, and CEO of The 831 Group; Michael Drumm, past fire chief of Markham, Illinois, and visiting assistant professor of the public services graduate program at DePaul University; Harold Cohen, deputy fire chief of Baltimore County, Maryland; and Bill Lowe, fire captain in Clayton County, Georgia.

Lowe cited "commitment to lifelong learning, enhance personal interest and passion of the fire service," while Wakeham wanted "to be the best I could be, to take on a challenge, to climb another mountain."

Their views were echoed by Cohen and Drumm. "This degree would assure that my professional journey would be lifelong, offering me several roads to travel, paths to explore and paths to blaze," says Cohen. According to Drumm, the "responsibility . . . to bring the most complete and up-to-date administrative skill possible to my department and the fire service" was important.

As for the benefit of such degrees to the fire service, Wakeham believes that "research skills help provide the body of knowledge needed. Leadership gets better, and those we deal with in everyday settings may look at us and the entire service much differently."

Lowe says that the benefit to firefighters comes from having "fire service practitioners with the universally recognized expertise as researchers and educators.... The visibility and contributions of the fire service's doctoral pioneers raise the academic bar."

In the twentieth century, law enforcement achieved professional status largely because of its acceptance by the academic community as an occupation worthy of study at the doctoral level. The academic infrastructure for the police service has been created and is producing the doctrine for the discipline.

That's the next step for the fire service. "The fire service is breaking stereotypical professional paradigms," says Cohen, "allowing us to be welcomed into other professional societies.... It gives us the intellectual union card."

The fire service has made a good first step at the bottom of the academic ladder. Some fire service doctoral pioneers are reaching the top academic rung in other disciplines, and such doctorates should be sought. But surely the fire-service discipline needs at least one doctoral program in fire-service studies.

To compete in the twenty-first century, the fire service needs bold action at the top of the academic ladder. "The leaders of today's fire service are at the forefront of the ever-changing definition of fire service," Drumm says. He notes that leaders are "responsible for setting the vision and challenging the members of the fire service to make the vision a reality."

The next step?

What needs to be done to achieve the vision of a fire-service studies doctoral education infrastructure? Fire service studies doctoral programs would need to be developed in universities that can involve faculties who bring terminal degrees, or PhDs, says Dr. Sandy Bogucki, associate professor at Yale University of Medicine, Section of Emergency Medicine. She's also the medical director for the New Haven and Branford, Connecticut, fire departments, as well as Branford's fire surgeon.

Ideally, faculty would have degrees in structural engineering, fire-protection engineering, fire and hazmat chemistry, public administration, decision science, medical/EMS, and criminal justice. Those faculty members would have to commit to taking the new program through the academic process to have it recognized by the university as a program of studies. Those awarded the degree would generally remain in academia, and their careers would be devoted to advancing the science and methodologies of the fire-service discipline with full specialty preparation and advanced research techniques now in the tool box.

Bogucki sees a "push/pull" fire service academic infrastructure evolving. The academic requirements for firefighters and fire officers is going up, so the "pull" to learn the science and research behind the discipline's doctrine is increasing. For example, all candidates who sit for the Fire Department of New York firefighter entrance exam are required to have thirty college credits. In 2007, all candidates for the FDNY battalion chief exam will be required to have a bachelor's degree. In 2009, the academic requirement for the EFOP and the CFOD will be a bachelor's degree.

"The push is that a certain percentage of bright, highly motivated individuals who get their degrees for professional qualification will be hooked on the academic aspects of the fire service, wanting to pursue higher degrees and answer the questions that arise during operations by conducting research using rigorous scientific methodologies," says Bogucki. "That's the push the fire service can create toward doctoral-level fire service research and teaching."

According to Chief Ronny J. Coleman, former California state fire marshal and founding chair of the Commission on Fire Accreditation, the fire service must address three needs to create a doctoral fire service studies infrastructure:

"First, a fire-service body of knowledge worthy of study at the doctoral level needs to exist. Second, broad populations of fire-service personnel with bachelor's and master's degrees eligible for and interested in advanced graduate study and research are needed, a critical mass. Third, we need a place for the new doctors of fire-service studies to practice and make significant contributions to the fire-service discipline."

Coleman compared the professional evolution for the fire service to surgeons. Many years ago, the local surgeon was also the town barber, hence the red-and-white pole. Surgery was practiced as a trade until science and research were applied to it, enforcing advanced study and certification to the discipline.

Coleman and Bogucki came to the same conclusion as Shaw in the nineteenth century: The fire service is like other science- and research-based professions. It needs to be studied and practiced at the highest academic and professional levels. And the fire service, with the help of academia, is responsible for creating the infrastructure to achieve that vision.

Shaw told us the fire service needs "precision study and training, as other professions do," as the fire-service leaders at Wingspread told us. Some fire service doctoral pioneers are leading the way, and academic requirements for firefighters and fire officers are increasing.

Today's fire-service leaders have described what needs to be done. In the twenty-first century, we must continue the journey by creating a doctoral fire-service studies infrastructure. This effort will help us better serve humankind.

Keep in mind that the degrees these scholars received are not in fire-service studies. Their degrees are in public administration, education, engineering, and psychology. They applied the doctrine of these disciplines to the fire-service environment in an effort to advance their doctrine. The secondary effect of the research is that the information may be valuable to our fire-service doctrine.

Fire service-related dissertations, 2002

The relation between burnout and compassion fatigue in firefighter-paramedics, Jennifer L. Bissett, Ph.D., University of Houston

Motivational factors and personality traits of individuals who decide to enter a career as a firefighter/paramedic, Robert D. Holborn, Ed.D., University of Central Florida

Regionalization of fire protection and emergency medical aid services: A comparative case study analysis of economic and social-political impacts, Brian Seichi Nakamura, D.P.A., University of Southern California

Decision-making in the public sector: A collective case study of fire sprinklers in Illinois' public university dormitories, Randal David Miller, D.P.A., University of Illinois at Springfield

Air quality impact evaluation of a hypothetical fire-fighting facility, William Glenn Fuller, D.E., Louisiana Tech University

A critical analysis of the fire accreditation process to discover if it impacts the effectiveness of paid, public fire departments, Ray O. Shackelford, D.P.A., University of La Verne

Fighting more than fire: Boredom proneness, workload stress, and underemployment among urban firefighters, John David Watt, Ph.D., Kansas State University

Getting a Doctoral Degree
2005

THE CONCEPTS OF *doctoral education* and *firefighter* are not usually connected in the same thought by anyone in the fire service, higher education or society in general. At the beginning of the twentieth century, *firefighter* and *bachelor degree* were not thought of together.

Today, there are thirty-two bachelor's degree and six master's degree programs related to fire service. In 2009, a bachelor's degree will be required to apply for the International Association of Fire Chiefs Chief Fire Officer designation, and the bachelor's degree will be the minimal academic requirement to get into the National Fire Academy Executive Fire Officer program. Finally, in 2007, you will need a bachelor degree to take the test for battalion chief in the FDNY.

Why should the firefighters reading this article be interested in doctoral education, especially if the idea of more school turns your stomach? Two words you did not learn in firefighter rookie school or any course you have ever taken: *doctrine* and *doctorate*.

Doctrine impacts you every day. Fire-service doctrine is

what we know, teach, and do. All of our national standards, training materials and books, plus your local standard operating procedures, manuals, drills, and traditions make up our fire-service doctrine. Each profession has its unique doctrine. What makes each one a profession is that all the people who practice in that field are following the same doctrine. A profession's doctrine is created through scientific research and disseminated through writing, teaching, learning, and application.

Today, fire-service doctrine is largely based on experience and consensus, not science and research. This explains why East Coast and West Coast fire departments practice differently. Why volunteer fire departments practice differently than paid fire departments. Why some apparatus responds with one firefighter and others respond with four firefighters. Most often, our books are based on the author's experience, which usually comes from one fire department. Our training manuals are based on our consensus standards. Groups of fire-service practitioners come together and agree on the content of the standards and training manuals, based on their collective experience, but that process is not based on research and science. Who is in charge of fire-service science and research? As yet, no one!

Now let's look at the word *doctorate*. According to the Carnegie Foundation study on the doctorate, the purpose of the doctoral education "is to educate and prepare those to whom we can entrust the vigor, quality, and integrity of the field. This person is a scholar first and foremost, in the fullest sense of the term. Such a leader has developed the habits of mind and ability to do three things well: creatively generate new knowledge, critically conserve valuable and useful ideas, and responsibly transform those understandings through writing, teaching and application. We call such a person a 'steward of the discipline.'"

The existence of a doctoral degree in a discrete discipline

is one of the most significant conditions that must exist for a discipline to be accepted as a true profession by the academic community, other professional disciplines, and society in general. The doctoral degree higher-education infrastructure is what generates the doctrine for the profession to use. This doctrine must be based on science and research, making it generalizable to the entire profession. The stewards of the discipline, those with doctorates, create the doctrine for the profession, by employing science and research.

There are fire-service practitioners today who have doctorates, but their degrees are not in fire-service studies because no such program exists. These individuals wanted the science and research skills but had to study other disciplines because studying the fire-service discipline at the doctoral level was not available to them. They studied fire-service-related disciplines like education, administration, sociology, psychology, business, engineering, medicine, history, and law. These practitioners realized that the fire service is an inter-disciplinary subject. We use the science and research for other disciplines in our daily practice.

When doctrine of these related disciplines is used in the unique fire-service environment, the various scientific laws and principles may be affected. For example, the reaction of a chemical in the laboratory is predictable. When a tanker truck full of the same chemical crashes into a building downtown during rush hour, the science related to the chemical is different than the laboratory science. Another example: The science and research related to labor management typically does not take into account employees who work twenty-four-hour shifts and sleep on the job, nor does it study a workforce made up of eighty percent voluntary labor expected to risk their lives doing the job.

These types of questions should be of interest to the fire-service discipline. The place to study such issues is in a fire-service studies doctoral program. Until we as a discipline can point to the science and research to justify our practice, we will continue to be subject to the whims of others who control our funding, standards, effectiveness, and efficiencies.

Consider that there are twenty-nine doctoral programs related to the police service. This higher education infrastructure generates about one hundred dissertations per year compared to about ten fire service dissertations per year. At Harvard University's Kennedy School of Government, there are twelve faculty, fellows, and researchers conducting research projects related to the police service. There are no programs listed related to the fire service. Is it any wonder that the police service receives more political and economic support? There are five doctoral programs related to the study of folklore. The fire service needs at least one doctoral program related to our discipline.

Getting a doctorate is not for everyone in the fire service, nor does every fire chief need a doctorate to be successful, but some firefighters reading this will be motivated by the idea of conducting research that can help the fire-service discipline become a true profession. A fire-service doctoral studies infrastructure needs to be created so the stewards of the discipline can have a place to conduct their research, from which we all can benefit. This connection among science, research, and practice will help us all better serve each other and our communities.

What do you do with a doctorate? Those who graduate with a doctorate will assume three duties. First, you will generate new doctrine for the fire service to use. You will do this through scientific research. The questions you ask and the answers you generate will meet the highest possible standards of scholarship. Your work will be for the betterment of the entire fire service.

Second, your work will be based on a complete understanding of the history of the fire service and the foundations of the discipline. Your work will maintain the continuity, stability, and vitality of the fire service by continually investigating which ideas to keep and which ideas to reject. Through your work, the best of the past will be preserved for those who follow. You will respect other disciplines and continually look for relationships between the fire service and other fields for their mutual benefit.

Third, you will represent and communicate fire-service doctrine to the widest possible audience—from firefighter to policy maker, from lay public to Nobel laureates. You will do this through your writing, teaching, discourse, and practice. This duty requires leadership that exposes your beliefs, values, and biases to public scrutiny and review. Your work will be defendable based on the best available science and research.

If improving and impacting the overall fire-service discipline sounds interesting to you; if working in the national and international fire service arena is what you want; if the role of change agent, researcher, educator, philosopher, and advocate appeals to you; and if the concept of building up the fire-service discipline as a whole seems like a meaningful life goal, going for the gold tassel and becoming a steward of the fire-service discipline may be for you.

If you are still not convinced that doing all that work to get your doctorate will be worth it, think about this. When the first fire-service studies doctoral program is created, it will need textbooks and faculty. To teach in a doctoral program, you need a doctorate; to write books for a doctoral program, you need a doctorate. So, those firefighters who have doctorates are guaranteed part-time work or a new career in academia after they retire. Finally, people will call you "Doctor," and your mother will be proud.

Professor Frank Brannigan Taught Us More Than Building Construction
2006

PROFESSOR FRANK BRANNIGAN'S contribution to the education of the fire service as related to building construction doctrine is profound, legendary, and one of a kind. His work in this field will continue through his book, articles, video programs, and the tens of thousand of firefighters who have heard him speak on building construction. For the next hundred years when someone tells a rookie firefighter, "The building is your enemy. Know your enemy" and then hands the firefighter a *Building Construction for the Fire Service* textbook, Frank will smile. By reading this book the firefighter will have lifesaving knowledge.

Frank's legacy to the fire service and to each of his students transcends building construction. The lasting effects on firefighter safety that he made can only be enhanced by looking deeper into what he gave us. By discussing his work, we add to our understanding and appreciation of "The Ol' Professor." I know of no better way to honor him.

I am one of the lucky firefighters who had Professor Frank Brannigan as my teacher in four of my five fire science courses from 1974 to 1976. I was a young hotshot know-it-all-firefighter from Prince George's County, Maryland, Company 33 Kentland, Company 10 Laurel, and a Washington, D.C., firefighter who graduated number one in his rookie class. I was smart and fought a lot of fire; what could college teach me?

My college courses with Professor Brannigan contained important information that I applied immediately and continue to use today. But what I learned from Frank goes far beyond the fire-science subject matter. He was illuminating a philosophy for the fire-service discipline. The philosophical underpinnings of Professor Brannigan's classes are the lessons I feel compelled to share with you because they have shaped my study of contributions to and application of fire-service doctrine. Frank taught us what to learn, how to learn and why we should learn.

What

Frank's academic background was as an accountant. Having twelve accounting credits myself I immediately recognized Frank's logical, fact-based, A=L+P (Assets = Liabilities + Proprietorship) approach to teaching fire-service doctrine. Frank understood that real fire-service knowledge had to apply to all fire departments and all firefighting. He was not teaching the District of Columbia Fire Department way, or the Prince George's County Fire Department way, or the Montgomery County Fire Department way, or the New York City Fire Department way, or the Navy Fire Department way. Just as accounting doctrine must apply to all businesses, Frank introduced us to the idea that fire-service doctrine must be science- and research-based if it is to be generalizable to all fire departments.

To get our class to grasp this concept we started back at the beginning with vocabulary.

"You must know the meaning of the words. Every profession has its own vocabulary. Firemen [in the mid-seventies we were still firemen; we had not evolved to firefighters yet] go into burning buildings; an architect designed it; an engineer made the blueprints; then the carpenters, masons, electricians, and plumbers built it. We must know their language and what their words mean to us when the building is on fire."

The quotations I attribute to Frank may not be direct. After thirty years my ability for perfect recall is fading, and it would take too much time to find my class notes but I still have them.

This is just a small list of vocabulary we had to learn: *gravity, absolute, column, beam, truss, arch, force, compression, tension, shear, cantilever, load, dead, live, bearing, nonbearing, pre-tension, post-tension, concrete, mortar, cast iron, terracotta tile, metal deck roof fire, partial pressure of oxygen, consensus standard, ASTM, flame-spread rating, tunnel test, time temperature curve, exothermic, endothermic,* plus hundreds more. These words were not in any firefighting training class I had ever taken. Nor did any of my brother firefighters at the stations I hung out at in the seventies use these words around the kitchen table.

Most of you reading these words know the definitions, but at the time they were a foreign language to most of the fire service I was associated with. The fact that I and those around me did not know the concepts associated with these words made two things clear to me. First, experienced firefighters and officers could not teach me everything I need to learn. Second, I could not rely on my experience to teach me all I needed to learn. To this day the fire service relies on OJT (on-the-job training) as its principle teaching methodology. Frank was teaching us that experience alone did not work in the past, and it will not work in the future.

In 1975, my lieutenant did not understand the inherent danger of cutting a hole in a terracotta tile floor. I realized that being an officer did not mean they knew everything they should know. In 2003, a deputy chief (four trumpets) tried to countermand my order (I was an assistant chief with three trumpets) on not sending firefighters into a large grocery store with a metal deck roof, sitting on steel trusses, with the fire going through the roof. The number of trumpets you have does not directly correlate to what you know. We still need to teach building construction to all fire-service personnel regardless of rank. "Buildings don't kill firemen in new ways."

Thanks to Professor Frank Brannigan, the fire service has learned a lot of definitions from other disciplines. But the fire service as a discipline has a long way to go to become a true profession because we have so many different definitions for the same things. If you put ten firefighters from different states in a room and ask them to write the definition of *engine company* you will get ten different answers. Frank's guidance to us—"You must know the meaning of the words"—is still true today. It is a simple fact that all science begins with defining the words, and all professions have a common language.

Professor Brannigan taught us how to be fire-service professionals and competent firefighters in our departments. He made us realize that what we learned today may have to be revisited in the future because changes in fire-service knowledge and in other disciplines require it. Professionals must rely on science and research, not just how your fire department teaches you to do things. We must understand other disciplines, because their doctrine has a profound life-and-death impact on the fire service.

How

The Montgomery County College fire-science program was designed for firefighters. Every class was conducted twice, once during the daytime on one day and once at nighttime on another day. This made it easy for career firefighters on shift work or volunteers working regular hours to attend. If you missed a class Frank audiotaped his lectures and put them in the library with the slides. You were expected to go watch and listen at the library to get credit for attendance. These ideas are common today, but at the time they were revolutionary and made college available to many firefighters.

Frank made us read books, magazines, reference manuals (ASTM, UL, NFPA), take lecturer notes, do case studies, memorize, and do field work. He made us look up references and citations because "You can not understand the authors if you do not understand their sources." What a radical idea, especially for undergraduate associate-level learning.

For example: What does it mean for a wall and ceiling assembly to have a one-hour or two-hour fire rating? Frank made us understand ASTM E119, which are the test standards used to achieve building construction and materials fire rating. The wall and ceiling installation are perfect in the test conditions—e.g., nailing/screwing (number, position, and depth); joint taping; mudding; and no poke-throughs for electrical, plumbing, and HVAC. Rarely is the installation in the field perfect; Frank's analogy was, "A condom with a small pin hole gives no protection."

In 1980, during the final inspection of my new home, I noticed that the sheetrock in the attached garage was not taped at the joints, nail heads were not mudded, and the corners were not spackled. I informed the contractor that this installation did not meet the fire code. His response was that it did meet the code because the sheetrock had a two-hour fire rating stenciled on it. I said that is not enough. He replied that the building inspector had passed the house for occupancy. I said get an inspector here. The inspector who passed my home arrived. He also referred to the stenciling on the sheetrock. He said the building code calls for two-hour-rated sheetrock. I said that is correct, but are you aware of ASTM E119, which explains the installation standards? He was not. He called back to the office, where they looked up the standard and discovered I was correct. The building official ordered my contractor to fix my garage. Reluctantly the builder said, "I will fix this one (meaning my garage) but I am not fixing the rest of the development (about one hundred homes) because you approved them."

"Builders and building code officials are more concerned with the building falling down than with the building burning down." The fire service has to teach other disciplines what they need to know when it comes to fire protection.

For extra credit in Frank's classes we could attend the Johns Hopkins University Fire Service Colloquiums, held at the JHU Allied Physics Laboratory, then write papers about the topics presented. This was my first exposure to speakers with doctorates talking about fire-service issues from a scientific, experimental-research point of view. I had to look up a lot of words from those lectures. Before I went to the conference I had to look up the word *colloquium*. When I received my doctorate in 1990, Frank attended my graduation party. Their gift to me was an autographed copy of the second edition of *Building Construction for the Fire Service*. The inscription reads, "Burton, the best friends are old friends."

Frank understood a basic concept of epistemology: There is no practice without theory; there is no theory without practice. If you ever met Frank, he most likely had a camera with him. At his lectures you saw hundreds of slides (I am in one of those slides pointing to the hole that I fell through while searching a second floor) showing real buildings, real fires, and real or potential consequences. The pictures were from communities across the country because buildings burn the same way and gravity is the same from New York to Los Angeles to Chicago to Dallas and every place in-between. So the theory of building construction in the fire service had to apply to the practice of firefighting in all fifty states. In true professions there must be theory to justify practice, and the practice must be based on theory, if what we know and do is to be considered credible.

Each of Frank's slides is worth a thousand words. When you understand the theories related to the individual picture you can apply them to any picture or real building you encounter. That knowledge and ability is priceless and it may save your life. Those 35mm carousel slide trays and the stories (*Building Construction for the Fire Service*, third edition, pages 178–179, is my story) that went with each slide were teaching us more than building construction. Frank was demonstrating that professionals understand the theory behind their practice, and they do not practice anything that is not based on acceptable theory. "Gravity is an absolute. One brick can kill you just as easy as the entire wall."

Frank taught us how to learn by looking behind the page and the picture to see what the words, images, ideas, facts, principles, and concepts were based on by reading, writing, listening, experimenting, and experiencing. You must "undress the building" to see what is holding it up. We must undress our fire-service doctrine to see what it is based on.

Frank's students wanted to go to his classes because they were fun, interesting, entertaining, challenging, thought-provoking, and useful. All professors should strive for that level of teaching perfection.

Why

The why part of this article is the most subjective; I claim no special insight to Professor Frank Brannigan's motivation for his sixty-plus years of dedication to the fire service. My observations are based on being his student, friend, and colleague over the past thirty-plus years.

Frank was trying to help his students become leaders in the fire service, by helping develop individuals who were capable of leading the entire fire service to a new understanding. He knew that true leadership is about taking the fire service in a new direction by generating new knowledge for all to use in the advancement of the discipline. "If you want to get ahead, pick a topic no one is interested in and become an expert in it. Soon people will seek out your advice, regardless of your age or rank or department, which will give you leadership opportunities." Technically, in leadership textbooks, this is called "Expert Power."

People follow experts because they think the experts knows what they are talking about. Frank was the undisputed building-construction expert in the fire service. More importantly, the way he practiced his expertise should inspire all of us. So if you want to be the Frank Brannigan-like expert in your topic you will need six characteristics. Talking fast and wearing bolo ties are not on this list but may be important. You must be passionate about your topic. You must be committed to giving your knowledge away at little or no cost. You must make it understandable to all. You must make it useful to all. You must continue to learn more about what you are already the number-one expert on. Finally, you must have the courage to tell the fire service when it must change—like no standing on or standing under lightweight wood-truss roofs when there is a fire below it. Or, how unprofessional and dangerous it is not to wear your seatbelt in the firetruck.

My article on Brian Hunton sparked Frank to write, "We Don't Use Seatbelts." To have Professor Frank Brannigan read and reference my work is truly a career high.

While Frank was creating his own legacy in the twentieth century, teaching us all about building construction, he was showing us how to be the leaders and legends of the twenty-first-century fire service. Frank's true legacy is the intellectual and professional investment he made in each one of us.

I have added something new to my biography: "He studied fire science at Montgomery College with Professor Frank Brannigan." Identifying the professor you studied with is most often reserved for doctoral-level study. Because the philosophical underpinning of the professor shapes the student's thinking, teaching, research, and practice throughout his or her career. When I combine what Professor Brannigan taught me with my academic studies and my thirty-six years in the fire-service community I come up with the following:

As a member of the fire-service discipline, what I know and what I do not know has life-and-death consequences for me and others. Therefore, a fire-service calling demands my highest level of professionalism.

If I know Frank, when we meet him again he will want to know what we have done with what he gave us. As an accountant, Professor Brannigan expects a good return on his investment. As a firefighter, Frank will measure the interest owed by the contributions we have made to increase the professionalism of the fire-service discipline. We all need to work hard to get an A in Professor Frank Brannigan's next class.

Speeches

What Matters
2002

TWENTY-NINE YEARS and twenty days ago, I had the privilege and honor to be in your seat. Today, being invited by Class 329 to address you is a greater privilege and higher honor. Thank you.

On November 10, 1972, 1 graduated from DCFD Recruit Class 249. Today, I can't remember what the speakers had to say, and thirty years from now you won't remember what I had to say. Why? Because the speech doesn't matter. What will matter to you thirty years from today will be your health, family, pension, and career satisfaction. It will take courage, commitment, and competency to have these blessings.

On the topic of your health, the fire service has always been and always will be dangerous. That is why only an elite few enter the discipline. Early in my career, my lieutenant told me that firefighters have to get killed because it's part of the job. I have rejected that notion for thirty years. Even after September 11 I still cannot accept that belief.

You must decide today if being injured or killed is part of the job. I pray that none of you are ever faced with a World

Trade Center event. Since that infamous day, which we all vicariously experienced, the fire service now has a new understanding of what is possible. Your job over the next thirty years is to act on the probable to ensure your safety, health, and survival.

And you can "*do the right thing.*" Maintain you physical fitness to reduce your risk of a heart attack, which is our number-one killer. Wear your seatbelt going to and returning from alarms to reduce your risk of vehicle crashes, which is our number-two killer. Use all your PPE and BSI to protect yourself from long-term illness. Follow the department's SOPs and your training doctrine. Because in every NIOSH Firefighter Line-of-Duty Death Report, the victims and others at the incident did no follow the SOPs and training doctrine. It is simple, but not always easy to "*do the right thing*" when others around you are not. It will take *courage*.

From today on, you will have two families, the one you live with and the one you work with. Few disciplines share the unique camaraderie of the fire service. As a D.C. firefighter, you have passport and open invitation to any fire station in the world. Along with this is instant friendship and professional credibility. But the fire service cannot take the place of a loving spouse, children, and grandchildren who always want you to come home to them. Your challenge over the next thirty years will be to keep two lovers. Not an easy task, but firefighters are trained to do the impossible. You will know if the two passions have been satisfied at your retirement party if an equal number of your fire-service colleagues attend out of respect for the contributions you have made to the discipline, and family members attend out of appreciation of your love. You will have cared for your two families well. It will take *commitment*.

On the topic of your pension I will defer to your capable union leadership, benevolent government, and grateful community.

Now is the difficult topic of your career satisfaction. Each one of you will have to develop your own yardstick over the next thirty years to measure satisfaction. What will give you fulfillment and gratification? What will meet your needs, desires, and appetite? How will that change over time? The DCFD and the fire-service discipline are truly a calling that offers unlimited opportunity to test you mentally, physically, and emotionally. I cannot give you a measuring device, but I will share with you four ideals that have helped guide me over the past thirty years.

First, the art and science of being a firefighter is a life-and-death vocation that requires you to perform every task one hundred percent correctly, one hundred percent of the time. Because over the next thirty years, you do not know which task, at what time, could be the failure that changes a mistake into an accident that results in a tragedy.

Second, society trusts firefighters. It is your duty to act professionally, follow a code of ethics, and deliver the highest standard of care. What will help you reach this goal is to treat all people as your family, friends, or neighbors.

Third, do your best all the time. Then determine how you can do better the next time. This will require life-long education, training, practice, and research.

Fourth, honor the past. A lot of people and events got you to where you are today. Celebrate the precious present. You have accomplished much, and tomorrow belongs to no one. Believe in the future. You are responsible for creating it, and the future is where you will spend the rest of your life. It will take *competence*.

What will matter thirty years from now? I hope you have your health, a loving family, and career satisfaction. The price for these blessings will be thirty years of courage, commitment, and competence.

In conclusion, I have a favor to ask. I want to be added to your retirement party list. I want to see how you turned out, and I want to ask you if I was right about *"what matters."*

See the Light. Be the Light
2004

SEE THE LIGHT. Be the light. This is supposed to be a motivational, educational, funny, emotional, and memorable speech, all in forty-five minutes. How is that possible? It is possible because the speech is about you, the speech is about me, the speech is about all of us. The speech is about people in our history and about people in our future. It's about the people and events around us that can change our lives forever. I am not going to talk about anything new but it may help you remember things you already know. And I get to achieve one of my career goals of being a keynote speaker at FDIC. I highly recommend that you add that one to your life to do list. I would like to dedicate this speech to my wife Carolyn Smith-Clark, for she is the light of my life.

See the light. Be the light. The idea for the speech comes from the Bible story about Saul on the road to Damascus. Saul was a good government bureaucrat. His job was to go round up Christians and deport them or crucify them. He was so good at his job that his nickname was Saul the Persecutor.

While leading his troops to Damascus, just outside the

city, a bright light from Heaven struck Saul; he was knocked from his horse and blinded. A voice spoke to him: "Saul, Saul, why do you persecute me?" Saul yelled, "Who are you, Lord?" The voice replied, "Go into the city and I will be revealed to you." Saul had to be led into the city where he lay in bed blind and starving for three days. A Christian named Ananias was sent to him. Ananias laid his hands on Saul; the scabs fell from his eye, he saw again, and he was baptized a Christian. Saul took the name *Paul* and became a disciple of Christ. Paul is credited with a large part of the New Testament and he is seen as being the light that spread Christianity. See the light. Be the light.

Another title for this speech may be, "Do you see the light in the people and events around you? Are you the light for those you love?"

None of us will have experiences like Paul or do work that profoundly impacts the world like he did. The closest we may ever come is Bruno and Brenner unplugged at FDIC.

There is a psychological aspect to this speech also. Dr. Morris Massey tells us that who we are at the core-value level is basically fixed in stone by about the age of ten. But a significant emotional event can change us at any time in our life.

Now that I have set the theoretical foundation for the speech, I will tell you the topics. I am going to give you some examples from history that demonstrate how one person can make a big difference. Then I will share with you some people and events in my life that changed me forever. I hope that by the end of the speech you will appreciate your life and the lives of the others around you, and the experiences you have together will be seen with greater illumination. Let's begin.

In the summer of 1919 the Nineteenth Amendment to the Constitution was up for ratification. This change in our constitution would give women the right to vote in national elections for the first time in history. In twenty-six countries, including Russia and Germany, women already had the right to vote. My maternal grandmother Ethel was a suffragette. The women's suffrage movement had been trying to win the right to vote for eighty years. It was such a powerful force that President Lincoln asked the women to stop their work while the men fought the Civil War.

For the Nineteenth Amendment to pass, two-thirds of the states had to ratify it. The women needed one more state to pass it. Two states were left to vote: Delaware and Tennessee. Between the two Delaware was the sure bet and Tennessee was doubtful. Delaware defeated the amendment. Eighty years of work seemed down the drain, but undaunted the women invaded Tennessee. Things looked bleak in the Tennessee House of Delegates. On every straw vote the amendment was defeated by one vote. Now the final vote was to be taken. The assembly hall was packed. The pro-suffrage group wore yellow ribbons in their lapels; the anti-suffrage group wore red roses.

Harry Burn was a first-term delegate, at twenty-seven the youngest member of the assembly. Harry wore a red rose. A roll-call vote was requested, and the delegates stood in alphabetic order and yelled *Yea* or *Nay*. The vote was going by party line. Harry Burn was called; he stood and said *Yea*. With that one simple word, an eighty-year quest had been reached and women were given the right to vote. At first no one realized what had happened or why. What no one knew was that Harry had a letter in his pocket that he had received that morning from his mother. It said, "Harry, be a good boy and give women the right to vote." Harry was in big trouble with his party and would have been run out of town tarred and feathered, but the next day he stood in the assembly and said, "A boy can never go wrong by doing what his mother wants, and my mother wanted me to give women the right to vote." See the light. Be the light.

Our next stop is Detroit in 1965. Ben is a fourteen-year-old inner-city kid. Today he would be referred to as an at-risk youth. He had three strikes already: single-parent home, living in the housing development, and poor school performance. On top of all that he had a terrible temper. "This sickness controlled me, making me totally irrational" is how Ben described it. He would fight, throw things, and lash out at the least provocation. One day on the way home from school he and his best friend got into a fight. Ben took a knife out of his pocket, opened the blade, and stabbed his friend in the stomach. He stepped back, looked down, and saw the blade lying on the ground. The knife had hit his friend's belt buckle and broken off. There was no blood; there was no wound. Ben realized he could have killed his friend. He ran home distraught, crying and pleading to his mother to take away his anger. What could he do to rid himself of this sickness?

His mother told him to pray about it, read a book every week, study in school, and join the high school Junior ROTC. Ben did what he was told. In 1969 he graduated number one in his class and JROTC commandant for the city of Detroit. With these achievements he was eligible for a scholarship to any military academy, but he chose to attend Yale University to study pre-med. From there he received his M.D. from the University of Michigan. He studied neurosurgery in Perth, Australia, where he saved the life of the fire chief by removing a brain tumor. At thirty-five, Dr. Benjamin S. Carson became the youngest chief of pediatric neurosurgery at Johns Hopkins Hospital. Dr. Carson's specialty is hemispherectomy, cutting children's brains in half to stop seizures. He was the first to lead a team that successfully separated twins conjoined at the head. Dr. Benjamin S. Carson. The "S" stands for Solomon. See the light. Be the light.

Our last historical moment takes place in Seattle, Washington, on January 5, 1995. A box alarm is put out for Mary Pang's Frozen Food. Heavy smoke is showing when the units arrive. Four attack lines with three men each go in the front door. Twelve firefighters are doing their job. The best job, the most important job, the most dangerous job. Some time into the fire a loud noise is heard. Lieutenant Gregory Shoemaker yells over his radio GET OUT, GET OUT, GET OUT. Eight firefighters tumble out windows and doors. Lieutenant Shoemaker and three brother firefighters fall into the basement when the floor collapses. Heroic rescue attempts are in vain. What told Lieutenant Shoemaker to yell "Get out"? We will never know, but eight firefighters are alive today because he did. See the light. Be the light.

There are hundreds of events in our lives that have impacted each of us in important ways, most of which we are not even aware. Few of us see how we impact others, and it is unlikely anyone will write about us. But that does not make your experiences and my experiences any less important or valuable. We can add value to the experiences by seeing their importance in shaping who we are and how we behave.

I joined the fire service in August 1970. It was a Saturday afternoon in Landover, Maryland. I had just moved there from New Jersey and didn't know anyone. My wife sent me up the street to the store. When I came out I noticed the doors were open at the Kentland Volunteer Fire Department across the street, so I walk in. No one was there except the Dalmatian dog. I later learned that his name was Haligan. A couple of minutes passed, and the firetrucks returned to the station. The guys got off in softball uniforms; they were just getting back from a game. We talked, they handed me a beer, and asked if I liked to play softball. I said sure, so I became a volunteer firemen on the spot. For the next couple of weeks I hung out at the firehouse a lot. I was between jobs, school did not start until September, and my wife worked. I had instant friends, and we played ball and drank beer.

I had been in the fire department for about two weeks but never actual got on the firetruck. One afternoon during the middle of the week I was at the station sitting in the TV room watching *The Gong Show*. A young guy came in dressed in an Army uniform and we introduced ourselves. He was Jimmy Pannetta, a member of the FD who had just come home on leave. Jimmy asked me if anyone had shown me anything. No, they just told me to get gear off the spare gear rack and get on the truck, I said. Jim offered to show me some stuff and we went to the apparatus bay. He proceeded to show me how to stand on the back step of the pumper: Hold on, bend you knees. Then he said he was going to show me how to be the layout man. He showed me the bundle of hose with a rope around it. He explained that the truck would stop at a fire hydrant; the officer would open the door and yell "Layout." I was to step off the back step, pull the rope and hose off the truck, wrap it around the fire hydrant, and yell "Go" to the driver. I got it—now I am a real firefighter. Back to *The Gong Show* we went.

It's not ten minutes later that all the bells and lights in the station begin to flash and blare. We've got something and we are going. I get my gear and jump on the back step. Four of us are on the engine: the paid driver, the paid sergeant, Jimmy Pannetta in the bucket, and me on the back step. We turn left out of the firehouse, go four blocks, and turn right into my apartment complex. Fire is blowing out of the third-floor windows of an apartment building.

Sure enough, the engine stops, the officer opens the door and yells "Layout," I step off, wrap the hose around the hydrant on the left, yell "Go." And I'm *done*. I have no idea what to do next. I am just standing next to the hydrant watching the fire. It seemed like forever for help to come but within two minutes the next pumper from Kentland arrived with only driver Wayne Ramsey. I did not recognize him and he did not know me. But he did know that I had no idea what I was doing. The steamer cap was not removed for the hydrant, the bundle of hose was still wrapped around the hydrant, and hydrant wrench was still hooked to the rope. The hydrant was on a grassy island in the parking lot so the pumper could nose right in front, only four or five feet away. Wayne knelt to untangle the hose and take the steamer cap off. He told me, "Get that big hose off the front bumper and bring it to me." I can follow directions. The hose was the big five-inch soft sleeve with the fifteen-pound brass coupling on the end of it. I grabbed the hose about six inches behind the coupling and pulled it and swung around to give it to Wayne. I gave it to him, all right. The coupling hit him in the side of the head. He went down like a rock. I realized that not only did I not know what I was doing, I was dangerous.

Wayne shook off the clubbing, got himself together, and said to me, "Stand there and don't move." For an hour I was frozen in place watching this amazing ballet of firefighters and equipment preformed. The longer I stood there the sicker and sicker I got in my gut; it was a terrible feeling I had never experienced before. I never wanted to feel that way again. I promised myself I would never not know what to do next again. There was more to being a firefighter than softball, beer, and *The Gong Show*.

I stated to learn every thing I could. I read books, took classes, and asked questions. Within six months I was a driver; within a year I was a training officer. Chief Massey Shaw of the London England Fire Brigade said, "Men who are pitch-forked into learning the business of a firemen only by attending fires must of necessity learn it imperfectly." He said that in 1873.

Two years latter I joined the District of Columbia Fire Department. Wayne and I became friends. I was like a sponge. I could not get enough information about the business of a firefighter. I graduated number one in my recruit class and attended my first FDIC in 1972. I was full of myself: DC Firefighter, Kentland firefighter; I was practicing the fire business a lot. We did a great job. I bought my first house in Laurel, Maryland, and joined the fire department.

In February 1974 Laurel had seven fire fatalities in one month from three fires. All of the victims—children and adults—were dead before the firetrucks left the station. I had that same familiar sick feeling in the pit of my stomach. All my knowledge, all my skill, all my macho, all the men and equipment in the fire department, could do nothing to save those people. There had to be a better way.

In the early seventies the smoke detector was new on the market. There was only one brand: Stradatol. It cost sixty bucks and the batteries were three bucks each. But they detected smoke early in a fire and gave people time to get out of their home—they could save lives. The Laurel Volunteer Fire Department went on a massive public education campaign to get people to buy and install smoke detectors. The program was so successful that the National Fire Prevention and Control Administration picked it as a model for its national smoke detector campaign. The DCFD learned I had helped my VFD with the program so the city assigned me to the fire chief's office to help with D.C.'s smoke-detector program. Washington, D.C., was the first metro city to have a retroactive smoke-detector law and citywide campaign. In 1978 I was detailed to the National Fire Academy to develop and deliver a smoke detector training program for fire services across the country. We have accomplished a lot with smoke detectors over the past twenty years.

In January 2001 a house fire in Oak Orchard, Delaware, killed eleven family members; three generations gone. The smoke detectors' batteries had been removed. When your family, your friends, and your neighbors go to sleep tonight will working smoke detectors be on guard? Do you see the light? Are you the light?

It is 1976 and I am in the kitchen of Engine 14 in the District of Columbia. I am arguing about the importance of firefighters going to college. "You don't need a college degree to ride the back step" is the opinion held by my colleagues. I countered with, "The more educated we get the fewer firefighters will get killed." My lieutenant responded, "Firefighters have to get killed; it is part of the job." I was livid, I was so angry, but you can't break bad on your lieutenant. I had to vent my anger. I was taking a writing class in collage so I went in the office and typed an essay on the topic. My professor thought it we pretty good and encouraged me to get it published. I did. *Fire Command* published "I Don't Want My Ears Burned" in July 1976. These are the key points: Do fire service personnel believe that being injured or killed is part of the job? What is the fire service's acceptable casualty level? The fire service will stop being the most hazardous occupation only when we its members want it to. Every time an accident occurs we must determine who or what is the cause.

The NIOSH reports do just that. It is no surprise that our morbidity and mortality is our own responsibility. In every LODD report the victim and or others were not following SOPs or training doctrine and in most cases an officer knew about it and did not fix the error before it led to an accident that resulted in a tragedy. Firefighting is dangerous; we must perform every task one hundred percent right one hundred percent of the time. Because you never know what little error, what little mistake, what little omission will be the weak link that breaks the chain.

Raise your hand if you wear your seatbelt all the time responding and returning from calls. One national survey said that only fifty-seven percent of us do. Fifty-seven percent cannot be a passing score for anything we do. We can change. In January of this year, Captain Bob Van Dyke of the Atlantic City FD Ladder 1 took the seatbelt pledge after his battalion chief asked him to. It was a fire department rule and it was the right thing to do. He stuck a job aid, a piece of duct tape with the word *BELT* written on it, to the dashboard of the cab of the truck to remind him to put his seatbelt on and tell his crew to put their seatbelts on. See the light. Be the light.

It is 1980, and I go to work at the NFA full time. I am in charge of the Public Fire Safety Education curricula. I get pretty full of myself again; people from around the country call me for advise. I think I even got a call from NFPA. Then one day a friend of mine calls. Joan Schermerhorn is a fire-safety educator in Vermont. She had taught a class of elementary kids fire-safety stuff, including what to do when the smoke detector goes off at night. "Crawl to the door, feel the door, if it is hot, do not open it." Joan had this smart kid in class who asked, "What do you mean by *hot?*" Joan realized she did not have an answer but she knew this government expert. She calls and asked "What's hot?" I did not have the answer, and no one could tell me.

I had participated in the California burn tests for residential smoke detectors and sprinklers. My friend Jim Feld, a fire protection engineering student at the University of Maryland, and I set up some full-scale burn tests to see if by feeling a door with your hand you could tell the temperature on the other side. Guess what? You can't. We had lethal temperatures at the five-foot level outside the bedroom door, and the surface temperature on the inside of the door never went above body temperature. Now for your fire-safety education lesson. Hold hands with the person on both sides of you. One feels warmer to you than the other. To the person who feels warm to you, say, "If you were a door I would not open you; I would use a second exit."

How many other things do we teach because we have always done it that way, or because the big city does it that way, or because the NFPA standards say to do it that way? Is the way based on science and research? Most of what we think we know is based on experience and consensus. As life and death discipline we must do better than that. Where does our science and research come from? Who studies us?

The first doctoral dissertation I ever read was "The Educational Effectiveness of Fire Demonstrations" by Homer P. Hopkins. It was written in 1966, but I did not find it until early 1981. There are about ten dissertations done on the fire service annually; there are about a hundred done on police service and criminology. There are twenty-nine universities with doctoral programs in criminology or criminal justice; there are no doctoral programs in fire service studies. Would you believe there are five PhD programs in folklore? Does anyone else see this as a problem? We will never be a true profession until the academic community accepts us as a discrete discipline based on science and research. Chief Massy Shaw said that the fire service needs to be studied and practiced like other science-based professions. Chief Shaw may ask us, "What has taken you so long to see the light?"

Next week the Congressional Fire Caucus Institute will hold its annual banquet in Washington, D.C. This is a big-deal. Held in the largest banquet room in the nation's capital, it welcomes two thousand fire service and political VIPs. Every year at this two-hundred-dollar-a-plate dinner, they give out an award to somebody who has done something outstanding and notable for the fire service. One year they gave it to a volunteer firefighter who got a fire pumper sent to Bosnia during its civil war. They also gave it to a leader of Kuwait after the first Gulf War for his work putting out the oil well fires and preventing an environmental disaster.

One year I got to attend with a free ticket. The award was given to Cory Snyder. Mr. Snyder had watched a television program about thermal imaging cameras, called his local fire department, and found out they did not have one, so he and his buddies went door to door collecting money. When Mr. Snyder came to the podium to receive his award, 1,999 people spontaneously rose from their seats to give him and his friends a standing ovation. One person could not rise. Me. I was getting that same familiar sick feeling in my gut: Something was wrong with this. Because you see, Cory Snyder was eleven years old, and his friends were younger. How can we as a professional discipline, whose vocation has life-and-death consequences, rely on children to raise money for the equipment we need to do our job? When was the last time the police department relied on donations?

For the first time in history the fire grants have given up access to funding like never before. But consider that the police service has been getting twelve billion dollars a year in federal funding even before 9/11. The police service understands that money is critical to efficient and effective operations. For example, if you put a group of police chiefs in one room and a group of fire chiefs in another and give them both the same problem to solve, the fire chiefs come out with a solution whereas the police chiefs come out with a grant proposal. We need to learn a new phrase, "Show me the money. We need it to do our job and keep you safe."

I have one more life-affirming experience to share with you. At the age of twelve I attended my first funeral. My great-grandmother Mattie Hammond died at age ninety. To this day I can still hear her voice in my head calling me Burtie. None of you gets to call me Burtie. Mattie Hammond had five children, six grandchildren, and ten great-grandchildren. Grandma Mattie was the center of attention when we all got together. At the age of thirty I met her again. I was going through some of my grandfather's papers after his death. I came across a poem that Grandma Mattie had written to him on the occasion of his retirement. When I read the poem I could hear Mattie's voice in my head, and I saw her in a new light. My mother told me that Mattie wrote lots of poetry and that my Great Aunt Pat had the whole collection. On the next visit I asked if I could read more of Mattie's poetry. I was given a typed manuscript of more than thirty poems that covered all types of events in our family. I even found one about me. I would like to share it with you.

A Feather
Great-grandson brought a feather
from a bluejay wing today.
I don't know why the bluejay
threw that one away.
But maybe left, as a token this gift,
to a little boy on this lovely day in summer,
that doesn't play with toys.
Burt tries his skill in casting
to catch a great big fish
and if he ever catches one
that is his dearest wish.

—Mattie Hammond

Two years ago my eight-year-old grandson Golden gave me a bluejay feather. What a story. But it does not end there. Three weeks ago my Great Aunt Pat died at the age of 101. I loved her. She was the last of Mattie's children. My mother gave me a case and said, "Aunt Pat wanted you to have these." When I opened the case there was a book of the original handwritten poems of my great-grandmother. I flipped through it and found the bluejay feather. I will put Golden's feather with it. See the light. Be the light.

Finally, a wise old man went down to the shore to think and write. One morning he got up early to walk on the beach. He saw what looked like a person dancing in the waves. As he got closer he saw that the young man wasn't dancing; he was throwing something into the water.

"Young man, what are you doing?" he asked.

"I am throwing starfish into the water."

"I can see that. The question is why?"

"Well the sun is coming up and the tide is going out. If I don't throw them back into the water they will die."

"Young man, don't you realize that there are miles and miles of beach and thousands of starfish? You can't possibly make a difference." The young man bent down, picked up a starfish, and threw it into the water. And he said, "It made a difference to that one."

The old man was stunned and did not know how to reply. He just walked off. He had a terrible day and restless night. The next morning he went down the to the beach and helped the young man throw starfish into the sea.

The moral to the story is that vision without action is merely a dream; action without vision just uses up time. Vision with action can change the world.

My final thought is this. If you pay attention to the people and events around you, you will *see* the light. If you love the people and experiences you have and strive to make things better, you will *be* the light.

Farewell to the National Fire Academy
2014

I have no ambition in this world but one, and that is to be a fireman. The position may, in the eyes of some, appear to be a lowly one; but we who know the work which the fireman has to do believe that his is a noble calling. There is an adage which says that, "Nothing can be destroyed except by fire." We strive to preserve from destruction the wealth of the world which is the product of the industry of men, necessary for the comfort of both the rich and the poor. We are defenders from fires of the art which has beautified the world, the product of the genius of men and the means of refinement of mankind. But, above all, our proudest endeavor is to save lives of men—the work of God Himself. Under the impulse of such thoughts, the nobility of the occupation thrills us and stimulates us to deeds of daring, even at the supreme sacrifice. Such considerations may not strike the average mind, but they are sufficient to fill to the limit our ambition in life and to make us serve the general purpose of human society.

—Chief Edward F. Croker, FDNY, circa 1910

Burton A. Clark

AFTER 44 YEARS of service, this I believe. As a member of the fire-service discipline, what I know and what I do not know has life-and-death consequences for me and others. Therefore, a fire service calling demands my highest level of professionalism.

Our proudest moment is when we accept the fact that fire prevention, smoke alarms, and fire sprinklers will save more lives and property from fire than we ever will.

Our noblest deeds of daring are measured by having the courage to do each task one hundred percent correctly one hundred percent of the time, especially when others around us are not, because lives are at stake.

Our supreme success is achieved when there is no injury, or death, or loss from fire.

As firefighters we honor the past because many people and events have put us here today; we celebrate the precious present because we have accomplished much and tomorrow belongs to no one; and we believe in the future because that is where we will spend the rest of our lives serving the people and achieving our purpose.

Thank God for firefighters. Thank you God for letting me be a firefighter. Class dismissed.

"I know I speak for all of the firefighters I have worked with, who were primarily committed to ensuring life was protected. It is because I believe this statement I am sure as many people as possible should read *I Can't Save You, But I'll Die Trying* to remind themselves of what they do wrong, and the when and the why of it when a life, predictably near expiration, is made safe because of what can be read as unsafe practices."

–**Dennis Smith**, founder, *Firehouse Magazine,*
bestselling author, *Report from Engine Co. 82*

"By showing us the cultural and personal side of safety behavior this book can be an important guide to leaders, managers, and ordinary citizens. It is not only a vivid account of firefighting but is much more in making us aware of our own thinking under crisis conditions and making us understand what those who deal with crisis face."

–**Edgar H. Schein**, Professor Emeritus,
MIT Sloan School of Management